AF789981

Field Informatics

Toru Ishida
Editor

Field Informatics

Kyoto University
Field Informatics Research Group

Springer

Editor
Toru Ishida
Department of Social Informatics
Kyoto University
Kyoto
Japan

ISBN 978-3-642-29005-3 ISBN 978-3-642-29006-0 (eBook)
DOI 10.1007/978-3-642-29006-0
Springer Heidelberg New York Dordrecht London

Library of Congress Control Number: 2012944566

ACM Codes: I.2, J.4

© Springer-Verlag Berlin Heidelberg 2012
This work is subject to copyright. All rights are reserved by the Publisher, whether the whole or part of the material is concerned, specifically the rights of translation, reprinting, reuse of illustrations, recitation, broadcasting, reproduction on microfilms or in any other physical way, and transmission or information storage and retrieval, electronic adaptation, computer software, or by similar or dissimilar methodology now known or hereafter developed. Exempted from this legal reservation are brief excerpts in connection with reviews or scholarly analysis or material supplied specifically for the purpose of being entered and executed on a computer system, for exclusive use by the purchaser of the work. Duplication of this publication or parts thereof is permitted only under the provisions of the Copyright Law of the Publisher's location, in its current version, and permission for use must always be obtained from Springer. Permissions for use may be obtained through RightsLink at the Copyright Clearance Center. Violations are liable to prosecution under the respective Copyright Law.
The use of general descriptive names, registered names, trademarks, service marks, etc. in this publication does not imply, even in the absence of a specific statement, that such names are exempt from the relevant protective laws and regulations and therefore free for general use.
While the advice and information in this book are believed to be true and accurate at the date of publication, neither the authors nor the editors nor the publisher can accept any legal responsibility for any errors or omissions that may be made. The publisher makes no warranty, express or implied, with respect to the material contained herein.

The image used on the book front cover was designed by Akio Okumura.

Printed on acid-free paper

Springer is part of Springer Science+Business Media (www.springer.com)

Preface

Field is defined here as "a spatiotemporal area that is difficult to grasp via any analytical and/or engineering approach due to the coexistence of various individuals and entities, which results in the unexpected occurrence of seemingly accidental events thus necessitating our continuing commitment and care" (Osamu Katai). *Field informatics* aims to provide solutions to various problems arising in the field from the standpoint of informatics.

The components of field informatics are *description*, *prediction*, *design*, and *transfer*, and the methods for these components range widely. Because field informatics is a new research area, it lacks a founding method. We consider that various methods of various origins will be applied to realize description, prediction, design, and transfer in the field. For example, let us consider the social goal of revitalizing hilly and mountainous areas amid increasing depopulation. Though all agree that creating a new business model in these areas is important, it is not easy to find an effective way of attracting people to hilly and mountainous areas. By using field informatics, however, situations can be *described* with remote sensing (Chap. 1), biologging (Chap. 2), human sensing (Chap. 3), and ethnography (Chap. 4); the effects of various options can be *predicted* with system dynamics (Chap. 5) and multiagent simulations (Chap. 6), and solutions can be *designed* using methods such as inclusive design (Chap. 7) and participatory design (Chap. 8). In addition, the experience gained can be *transferred* by case writing (Chap. 9) and outreach communication (Chap. 10).

Field informatics can be applied to *nature observation*, *social participation,* and *innovation.*

Field informatics for natural observation is an extension of fieldwork, and uses both theories and methodologies to analyze information from fields scientifically. Field informatics can generate key technologies such as remote sensing and biologging to solve global problems such as food shortages and environmental degradation. The role played by informatics in natural observation involves collecting and accumulating data, and verifying and utilizing them in accordance with the intended purpose.

Although many species are in danger of extinction, no effective countermeasures are being taken for economic reasons. Faced with such problems, field informatics tries to help endangered species and humans coexist, observing the biosphere with various technologies. The field, as a place of natural observation, is also a venue for the productive and social activities of people. Mutual learning in concert with the people living there enables us to interpret phenomena. Here, the field is a real society, in which a range of factors intertwine in complex fashion. For example, agriculture, forestry, fishing, and livestock industries are affected by the environment, and also social systems such as politics and economy. Their problems and improvements can only be understood by viewing the field as a system consisting of natural and social activities.

Field informatics for social participation is a place where there is inherent value and people who protect that value. The new science required here involves modeling the complex and diverse phenomena arising in the field and transferring the resulting model to practitioners. Therefore, field informatics transcribes the words and deeds resulting from the actions of people and involves analysis and generalization. The obtained model must be shared by practitioners in the field in an understandable form.

However, only a subset of information in the field can be detected, managed, and controlled. Therefore, if simply used by someone unaware of this limit, a misunderstanding may arise in that the world becomes expressed only in terms of the obtained information. Qualitative studies based on social sciences are to be applied in parallel with information technology. Furthermore, people in the field are forced to actively participate in dealing with accidental events that are hard to control. Therefore, information should be seen as dynamic and active rather than static and passive.

Field informatics for innovation involves designing and realizing an information system that contributes to the human community through its application in different fields. In other words, it creates innovation via the mutual learning of the designers and users of information technology. From this perspective, the field is a site where new narratives are generated with the support of technical feasibility in the near future rather than those for mere observation. Field informatics for innovation incorporates technological progress during the development period, and creates a scenario that can be realized in the near future.

Resistance to new technology is often observed in the field. Also, the limits of explanations, which are due to the complexity of information systems, make it difficult to establish communication between researchers and people in the field. For example, pursuing the cause of software failure sometimes reveals a trivial manmade mistake. It is not always the case that the field is an easy place for computer scientists to get involved. However, in the future, since a large portion of innovations will be born in the field, more attention must be paid to it.

This book targets graduate students seeking tools and approaches for natural observation, field workers engaging in activities on social participation, and researchers and engineers pursuing innovation. Field informatics can be of benefit to not only students or researchers but also to those in other positions. For example, individuals in charge of government can utilize the methods of field informatics to form consensus and develop activities. Nonprofit organizations can use field informatics to conduct social programs more effectively. Those with sites such as agriculture, forestry and fisheries, hospitals, and so on can use field informatics to understand and solve various problems.

The members of the field informatics research group in Kyoto University who contributed to the work represented in this book are as follows: Nobuaki Arai, Hiromitsu Hattori, Toru Ishida, Shigeo Matsubara, Kazuyuki Moriya, Yuu Nakajima, Junichi Okuyama, Tetsuro Sakai, and Takaaki Tsuji of the Department of Social Informatics; Osamu Katai of the Department of Systems Science; Yutaka Yamauchi of the Graduate School of Management; Hajime Kita and Yuichi Nakamura of the Academic Center for Computing and Media Studies; and Takayuki Shiose of the University Museum. Reiko Hishiyama of the Department of Industrial and Management Systems Engineering, Waseda University, also joined the research group. The specialties of the members range from informatics, engineering, agriculture, and sociology to pedagogy. The fields targeted by each member also differ, ranging from environment conservation to social education for international cooperation. Specifically, the research activities of the biosphere in Southeast Asia—large-scale traffic simulations, participatory "mono-zukuri (the spirit to produce excellent products)" workshops, inclusive design workshops, distance learning, and intercultural collaboration—are included. Our research group, with such practical research activities as its background, is consolidating the various methods described herein as field informatics and providing lectures open to the public.

Toru Ishida
July 1, 2012

Contents

Author Biographies

Nobuaki Arai is an Associate Professor in the Department of Social Informatics, Kyoto University. He holds a Doctor of Agriculture from Kyoto University. He launched the Southeast Asia Sea Turtle Associative Research (SEASTAR) project in 2000. In the SEASTAR project, he is promoting sea turtle migration research, dugong biological studies and Mekong giant catfish tracking, mainly in Thailand. He is a chairperson of the Japanese Society of Biologging Science.

Hiromitsu Hattori is an assistant professor in the Department of Social Informatics, Kyoto University. He was a Research Fellow of the Japan Society for the Promotion of Science from 2004 to 2007. From 2004 to 2005, he was a Visiting Researcher in the University of Liverpool, UK. In 2006, he was a Visiting Researcher in the Massachusetts Institute of Technology, USA. His main research interests include autonomous agents and multiagent systems. He has been working on multiagent-based simulation and human behavior modeling.

Reiko Hishiyama is a professor in the Graduate School of Creative Science and Engineering in Waseda University, where she directs the Intelligent Information System laboratory. She received her Doctor of Informatics in 2005 from Kyoto University in Japan. She taught at Kyoto Women's University, before joining the faculty of Waseda University. Her current research interests include artificial intelligence, autonomous multiagent systems, knowledge representation, autonomy-oriented computing and related areas. The topic is viewed as an interdisciplinary field where computer science intersects with organization science, sociology, psychology, and other fields.

Toru Ishida is a professor in the Department of Social Informatics, Kyoto University. Until 1993, he was a research scientist in NTT Laboratories. He spent some time at Columbia University, Technische Universität München, Université Pierre et

Marie Curie, the University of Maryland, Shanghai Jiao Tong University, Tsinghua University, and Hong Kong Baptist University as a visiting/invited scholar/professor. He was made an IEEE fellow in 2002 for contributions to autonomous agent and multiagent systems. He currently studies social informatics and is running research projects related to digital cities and intercultural collaboration.

Shigeo Matsubara is an associate professor in the Department of Social Informatics, Kyoto University. From 1992 to 2006, he was a research scientist in NTT Communication Science Laboratories, NTT. He received his Ph.D. degree in Informatics from Kyoto University. During 2002–2003, he was a visiting researcher at the University of California, Berkeley. He was also an advisor in the NICT Language Grid project from 2006 to 2007. His research focuses on multiagent systems and information economics. He has published in the Artificial Intelligence journal and other academic journals.

Kazuyuki Moriya has been a professor in the Graduate School of Informatics, Kyoto University since 1998. From 1982 to 1991, he was an assistant professor in the Faculty of Agriculture, Miyazaki University. He was an associate professor in the Faculty of Agriculture, Kyoto University from 1991 to 1998. He received his Ph.D. in Agriculture from Kyoto University in 1990. His major field was animal science, especially population genetics of domestic animals, but recently his research interests include agricultural system analysis, animal movement analysis using GPS and environmental learning. He is a vice-president of the Society of Beef Cattle Science, a director of the Japanese Agricultural System Society and a director of the Japanese Society of Animal Breeding and Genetics.

Yuu Nakajima is an assistant professor in the Department of Social Informatics, Kyoto University. He was a Research Fellow (DC1) of the Japan Society for the Promotion of Science from 2006 to 2009. From 2009 to 2010, he was a Researcher in the Department of Social Informatics, Kyoto University. He received his Ph.D. degree in Informatics from Kyoto University. His research interests include large-scale multiagent systems, agent modeling and multiagent-based simulations.

Yuichi Nakamura received his B.E. degree in 1985, and his M.E. and Ph.D. degrees in Electrical Engineering from Kyoto University in 1987 and 1992, respectively. He worked as an assistant professor in the University of Tsukuba from 1993 and then as an associate professor since 1999. He is currently a professor in Kyoto University. His research interests and activities include human–computer interactions, video analysis, and video utilization for knowledge sources.

Junichi Okuyama received his Ph.D. degree in Informatics from Kyoto University in 2007. He has been a GCOE (project) Assistant Professor at the Department of

Social Informatics, Kyoto University, since 2007. He is involved in research into aquatic animals using biologging and biotelemetry—the method of observing invisible animals—which uses data loggers and transmitters with multiple sensors to trace the activities of organisms. He conducts basic research into the preservation of endangered species such as sea turtles, and into coexistence between humans and such species.

Tetsuro Sakai is a professor in the Department of Social Informatics, Kyoto University. Until 1998, he was a professor in the Graduate School of Agriculture, Kyoto University. He received his Ph.D. in Agriculture from Kyoto University in 1986. His major field was forest science, especially forest management and planning. Recently his research interests include mountainous region issues and wooden cultural properties. He is a director of the Japanese Agricultural System Society and an executive director of the Japan Forest Engineering Society.

Takayuki Shiose got his doctoral degree from the Graduate School of Engineering, Kyoto University in 2004. He received master's and bachelor's degrees in Precision Engineering from Kyoto University in 1998 and 1996. From 2000 to 2002 he was an assistant professor in Kobe University, and from 2002 to 2008 an assistant professor in the Department of Systems Science, Graduate School of Informatics, Kyoto University. Currently, he is an associate professor in the Kyoto University Museum. His research interests include communication design and inclusive design. Academic awards received include the Best Paper Award at the 7th IEEE International Workshop on Robot and Human Communication (1999).

Takaaki Tsuji completed the Ph.D. program without dissertation in the Graduate School of Education at Kyoto University in March of 2007. After that he worked as a postdoctoral fellow in the Center for Human Activity Theory at Kansai University. From April 2008, he was a Global COE Assistant Professor in the Department of Social Informatics at Kyoto University. He is currently an associate professor and a vice director in the center for evaluation, Akita University. His research focuses on higher education and university evaluation. He has conducted research to analyze students' experiences in remote learning and cooperative learning via activity theory and situated learning theory. Recently, he has engaged in practical research on faculty development, and he is producing a series of video teaching materials on this topic.

Yutaka Yamauchi is a Senior Lecturer in Kyoto University, Graduate School of Management. Prior to joining GSM in September 2010, he was a researcher at Palo Alto Research Center (PARC), a Xerox Company. He obtained Master's and Bachelor's degrees in Computer Science from Kyoto University and a Ph.D. in Management from UCLA Anderson School. His research revolves around the question of what it means to know in organizations. Specifically, he has been studying knowledge sharing, learning, and change.

Chapter 1
Remote Sensing and Geographic Information Systems

Tetsuro Sakai

Abstract One of the most basic types of information used in field work is spatial information about a specific target area. Various types of information are spatial, including information concerning topography, land use, social infrastructure, climate, and manufacturing infrastructure. All of this information must be appropriately gathered in accordance with the objectives of a study. Remote sensing and geographic information systems (GIS) are among the many useful means for gathering and analyzing such spatial information. Using aerial photography and satellite images obtained through remote sensing, one can gather information covering wide geographic areas, such as information about natural resources or about the environment. For example, the distribution of vegetation can provide clues regarding the expansion of desertification or the state of food production. In addition, more detailed spatial information can be compiled by combining these methods with field work or existing data. Positioning data attached to spatial information ca.n then be analyzed using a geographic information system (GIS). A GIS is both a database of spatial information and a tool for its analysis. For example, GIS analysis of landform data or precipitation data may be used to predict natural disasters. In this chapter, we review these methods of gathering and analyzing data.

1.1 Remote Sensing

1.1.1 Overview of Remote Sensing

"Remote sensing" is a technical term coined during the Space Age of the 1960s. As the term implies, remote sensing techniques are used to analyze objects that are far

T. Sakai (✉)
Graduate School of Informatics, Kyoto University, Yoshida-Honmachi, Sakyo-ku,
Kyoto 606-8501, Japan
e-mail: sakai@i.kyoto-u.ac.jp

T. Ishida (ed.), *Field Informatics*, DOI 10.1007/978-3-642-29006-0_1,
© Springer-Verlag Berlin Heidelberg 2012

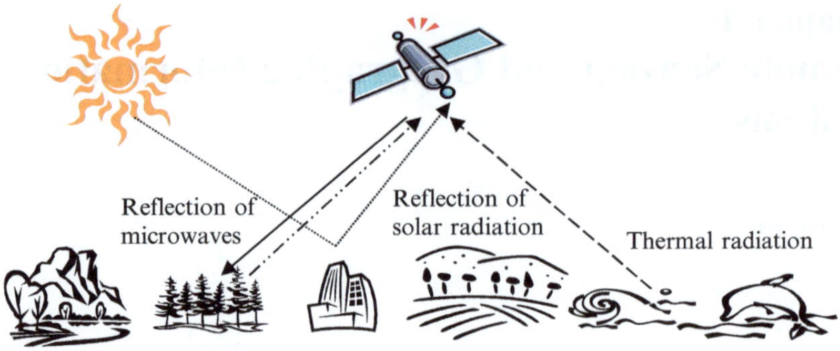

Fig. 1.1 Conceptual outline of remote sensing

away, such as to analyze what objects are or what states they are in. To identify the characteristics and status of a target object, the most commonly used remote sensing methods involve the reflection and radiation of electromagnetic waves. Target objects or phenomena can be deciphered and analyzed based on their unique electromagnetic wave characteristics, as summarized by the following: "All objects, if their types and environmental conditions differ, have different characteristics in terms of the reflection or emission of electromagnetic waves." Other remote sensing methods include the use of magnetic or gravitational force instead of electromagnetic waves.

Remote sensing, which covers wide-scale terrestrial, atmospheric, and oceanographic data collection as well as the monitoring of global-scale environmental shifts, has applications in diverse fields. In terrestrial science, remote sensing is used as a means of acquiring and analyzing data about the environment and natural resources, such as data on land use, land cover, vegetation changes, crop growth, and grain harvests. In oceanography, areas of remote sensing inquiry include sea level, water pollution, plankton distribution, and sea temperature, while in atmospheric science, remote sensing can be applied to examine the composition of minor atmospheric constituents, such as carbon dioxide and ozone, and to analyze cloud formations and other weather phenomena.

Figure 1.1 displays a conceptual diagram of remote sensing. There are three types of remote sensing. First, there is visible spectrum/reflection infrared remote sensing, which measures reflected sunlight. Second, there is thermal infrared remote sensing, which measures heat radiation emanating from objects. Third, there is microwave remote sensing, which measures the reflection of emitted microwaves. The type of remote sensing applied depends on the purpose of an investigation.

Information about the properties of a target object can be obtained by measuring reflected solar radiation using both visible spectrum and reflection infrared remote sensing and then by comparing the differences between the two measurements. The reflection of solar radiation contains electromagnetic waves in the form of ultraviolet light, visible light, and infrared light, with the peak of solar radiation coming in the form of visible light. Because the electromagnetic wavelengths being observed are short, and because electromagnetic observations make use of reflected solar radiation, this type of remote sensing is not possible at night or under cloud cover. Measurements are also affected by such observational conditions as scattering within, and reflection from, the atmosphere, the position of the sun at the time of observation, and topographic features on the earth's surface. Because of these effects, it is necessary to perform corrections for atmospheric and topographic conditions before using these images. An example of electromagnetic observations is those of cloud distribution conducted by Himawari, a weather observation satellite. Himawari provides a wealth of weather and climate data by measuring the reflections of not only the visible light spectrum, but also the infrared spectrum.

Thermal infrared remote sensing measures the thermal radiation emitted by a target object. All objects on the earth's surface emit thermal infrared radiation, and it is possible to estimate the temperature of land and ocean surfaces by measuring this thermal radiation. Measuring ocean surface temperature, for example, provides important information for global-scale weather observation, as well as for the fishing industry as a way of predicting good fishing grounds. In addition, remote sensing of the electromagnetic waves that emanate from objects is not limited to thermal infrared radiation. Night-time artificial light or lightning discharges can also be measured by remote sensing. Night-time artificial light can serve as an indicator of economic activity in a country or area or of the extent of damage in areas hit by earthquakes and other disasters.

Microwave remote sensing has two methods: the passive method, in which microwaves emitted by an object are measured directly, and the active method, in which fixed-wavelength microwaves are first emitted from the sensor, which then measures factors such as the strength of the resulting scattering (backscattering coefficient). The latter method is referred to as synthetic aperture radar (SAR). Using SAR, we can obtain information about a target object, such as its geometric shape or, by the Doppler Effect, its velocity. Because long-wavelength microwaves can penetrate clouds and particulate matter in the atmosphere, microwave remote sensing's main characteristic is its ability to make observations regardless of weather conditions or time of day. However, because microwave radiation is directional, microwave remote sensing is also strongly affected by ground topography.

Devices that measure electromagnetic waves reflected by or emitted from objects are called remote sensors, with two of the most familiar examples being digital cameras and scanners. Electromagnetic waves have four elements: frequency (wavelength), direction of propagation, vibration amplitude, and plane of polarization. Remote sensors are set up to measure some or all of these elements according to the investigative purpose. The efficiency (resolution) of a sensor used in remote sensing is expressed in terms of how large (wide) of a target area is covered by a single pixel. For example, a resolution of 30 m means that a single pixel covers an area of 30 m^2. Usually, high resolution refers to a few meters or less, medium resolution refers to several dozen meters, and low resolution refers to several hundred meters or more.

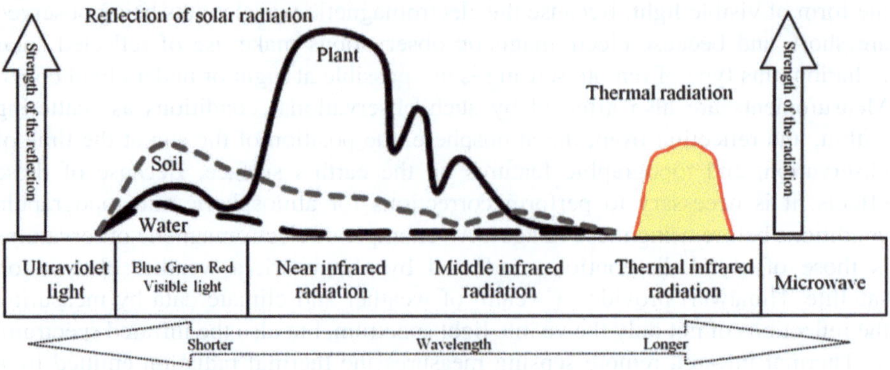

Fig. 1.2 Electromagnetic spectral bands and reflection/radiation characteristics

Every object has unique reflection and radiation characteristics in response to different frequencies of electromagnetic waves. These characteristics are called spectral reflection characteristics. Figure 1.2 illustrates the relationship between the electromagnetic spectral bands and these reflection and radiation characteristics. The horizontal axis represents the wavelengths of electromagnetic waves across each of the electromagnetic spectra. The vertical axis illustrates the strength of the corresponding reflection and thermal radiation for plants, soil, and water, respectively. Because plants reflect solar radiation with their leaves, their reflection characteristics are strongly affected by the reflection characteristics of chlorophyll. Chlorophyll absorbs comparatively more red light and blue light than green light and strongly reflects near-infrared range electromagnetic waves. The more active a plant is, the stronger its reflection tends to be. Using this reflection characteristic of plants, it is possible to observe the activity of vegetation. Soil reflection peaks in the visible light spectrum, growing weaker as it moves into the infrared spectrum. Reflection from water in the visible light spectrum is weaker than that from the earth's surface, and water barely reflects radiation at all in the infrared spectrum. To determine the composition of an object using the respective reflection characteristics of plants, earth, and water, the most commonly used sensors measure reflections in four spectral bands: the three colors of visible light—red, green, and blue (RGB)—and near-infrared light.

1.1.2 Methods of Processing Image Data

Remote sensing produces information in the form of digital data, which give the reflective brightness value for each pixel across all of the wavelength ranges measured by a sensor. The reflective brightness value, which is called a digital number (DN), is a one-byte (8-bit) integer value between 0 and 255 in most cases, although values can differ depending on the sensor. The wavelength range that a

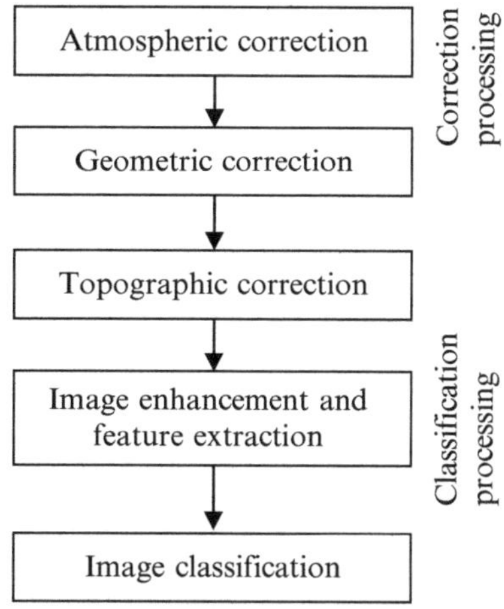

Fig. 1.3 The process of image data processing

sensor measures is called a band or a channel. The most commonly used multispectral sensors (MSS) measure the four bands mentioned above, namely, red, green, and blue light within the visible spectrum, and near-infrared radiation.

The processing of remote sensing image data, as shown in Fig. 1.3, includes two major areas: correction processing and classification processing. In general, after corrected sensor image data are obtained, atmospheric, geometric, and topographic corrections are performed. Next, the features of a target area or object are extracted and used to perform quantitative image classification.

Atmospheric correction is conducted to remove distortion caused by scattering that occurs in the atmosphere in the interval between solar radiation's reflection off the surface of the earth and its measurement by a sensor. Some simple methods of correction include those using values for an object whose reflection ratio is already known or those performing comparative calculations between spectral bands.

Geometric correction aims to correct for distortions in the sensor itself or distortions created by the method of projecting a map onto a two-dimensional surface. Such correction is normally conducted by comparing maps and images of the target area and designating several overlapping points as a basis for correction. Satellite image data contain information on the positions of an image's center point and its four corner points, and corrections are performed automatically based on the positions of these five points.

Topographic correction is conducted to correct for differences in the angle or angular direction (inclination) of the reflection of solar radiation caused by reflection off a slanted surface. Corrections are made based on the angle and inclination of the terrain, as calculated using digital topography maps (digital elevation models: DEM), and the position of the sun at the time of observation.

One simple method for making these corrections is to perform comparative calculations between spectral bands. Topographic correction is unnecessary when the target area is mostly flat.

The methods used for image enhancement and feature extraction are similar to those used to process photographs taken by digital cameras. Several methods are used for image enhancement processing. One method is color tone conversion, which uses a histogram (brightness frequency distribution) of reflective brightness values to convert color density. Another method is to make information from outside the visible spectrum visible through the use of color synthesis. Feature extraction includes such methods as spectral feature extraction, using principal component analysis or vegetation index, textural feature extraction, and spatial filtering, among others. Image enhancement and/or feature extraction can be important for grasping the features of a classification target in accordance with the study purpose.

Image classification begins by determining, according to the purpose of the analysis, the classification classes to be used for analysis. In other words, generalized classification classes for analysis are determined according to what one wants to know about a target area or what one hopes to discover. For example, if one wants to examine land-use conditions, classification classes such as rice paddies, fields, forests, buildings, and water could be useful. After determining the classes, one can then determine the variables to be used in the classification (classification variables) based on the features of the classes, which are derived using image enhancement and feature extraction. In general, the respective values across each spectral band and index, or characteristic quantities derived by performing calculations across spectral bands, are used as classification variables. It is also important, when choosing classification variables, to consider previous research. When a wealth of previous information exists about a target area, the method of classification can be based on (or "supervised" by) this information. This method is called "supervised classification." In contrast, when there is little previous information about an area, classification is conducted through statistical methods and is called "unsupervised classification."

In supervised classification, the locations appearing in images are specified in terms of each classification class and these locations are selected as training areas. Using data from these training areas, the statistical values of the classification classes are calculated and the entire target area is then classified using these statistical values. The maximum-likelihood classification method is generally used as the algorithm for classification. The accuracy of the classification can then be verified by comparison with the classification result of the training area. The characteristic features of this method are that classification classes are predetermined and that the accuracy of classification can be verified.

For unsupervised classification, first a fixed number of pixels is extracted by random sampling of the entire target area. Then, using the variables of those pixels, cluster analysis is used to classify the pixels into several classes before seeking the statistical values for these classification classes. Unsupervised classification then consists of applying these statistical values to classify the entire target area. For this method of classification, a further step is necessary, which involves manually defining the meaning of the classes based on a comparison between the different

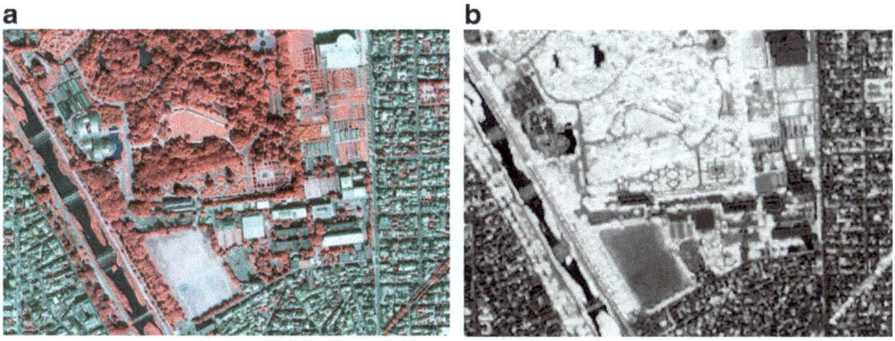

Fig. 1.4 A false-color image (**a**) and an NDVI image (**b**) based on an Ikonos satellite image showing the Kyoto Prefectural Botanical Garden and the surrounding area

types of land cover isolated by the automatic map analysis and independent observations (e.g., field surveys) of what they represent.

This type of image processing can easily be performed on a PC using commercially available remote sensing analysis software or free open-source software (such as MultiSpec © Purdue University). Most software products come equipped with the essential analytical and classification functions that allow users to obtain classification results simply by setting the parameters.

1.1.3 Vegetation Index and Land Cover

For land-cover classification of continental areas, the principal elements are plants, soil, and water. Plants are characterized by weak reflection in the blue and red bands but strong reflection in the near-infrared band. From these characteristics, the use of vegetation indices based on brightness values in the red and near-infrared bands was proposed. In the most commonly used vegetation index, called the normalized difference vegetation index or NDVI, higher index values indicate higher levels of vegetation activity. The NDVI is calculated using the following formula:

$$\mathrm{NDVI} = (\mathrm{NIR} - \mathrm{RED})/(\mathrm{NIR} + \mathrm{RED})$$

$$\mathrm{NIR} = \text{the brightness value of the near infrared band}$$

Figure 1.4 illustrates a false-color image and an NDVI image. False color refers to a method of transforming information from outside of the visible spectrum into something visible. When red is assigned to the brightness value for near-infrared light, green is assigned to that of visible-spectrum red light, and blue is assigned to that of visible-spectrum green light, vegetation appears in red (as illustrated in Fig. 1.4a). Figure 1.4b presents an NDVI image. The whiter areas represent higher NDVI values, indicating more active vegetation.

Some problems with satellite remote sensing include the very high cost of image data, the fact that imaging is affected by factors such as weather, the uncertainty of being able to obtain images not obscured by clouds, the fact that analysis results often vary depending on how the analysis is handled, and the existence of several formidable technical obstacles. Because of these problems, the issue of inadequate development, not only at the research level but also in terms of real business applications, requires future consideration. Needs-based, value-added information services, as well as the establishment of a standardized method of analysis, are necessary. Additionally, transmission techniques for acquiring images in real time, archives of past image data, and IT-based services will be more important in the future. Furthermore, the development of systems allowing decision-making or communications to be conducted easily through images and video is also valuable. Geographical Information Systems (GIS).

1.1.4 Overview of Geographical Information Systems

A geographic information system (GIS) is a system in which map information, along with various additional information, can be displayed and referenced using computers. GIS was originally developed in Canada for farmland revival and development in the 1960s. Use of GIS is increasingly widespread in all fields that handle space–time information thanks to advances in computer processing power and memory capacity and the refinement of computerized systems. Fields using GIS include those involved in the planning and management of urban living infrastructure (e.g., houses and roads), agricultural production facilities (e.g., fields), and the natural environment (e.g., forest protection), as well as military usage. A GIS handles information referenced in terms of space–time coordinate values. The *m*easurement of environmental variables, the *m*apping of features, the *m*onitoring of environmental changes, and the *m*odeling of plans or contingency plans are called the "four Ms" in GIS, and these four Ms plainly state the qualities of GIS.

In GIS, space–time information is managed according to units called layers. Figure 1.5 illustrates the concept of layers. While with maps, a wide array of information is simply laid out on a single sheet, GIS is made up of many separate sheets—called layers—representing different types of objects, such as rivers, roads, train lines, and the like, and these layers are managed in unison. The two representative data structures used in GIS are raster format and vector format. These data structures are conceptualized in Fig. 1.6. The vector format data structure encodes the coordinate values of a target object, turning them into a database composed of points, lines, and polygons. Although it is possible to determine coordinates or positions by setting the precision at the user's discretion, the computational algorithms involved are very complex. Data input also requires a large amount of time and effort. In contrast, the raster format data structure distributes a target object's variables for each factor into a grid corresponding to actual space. Although the computations involved are simple, the larger size of a raster bitmap

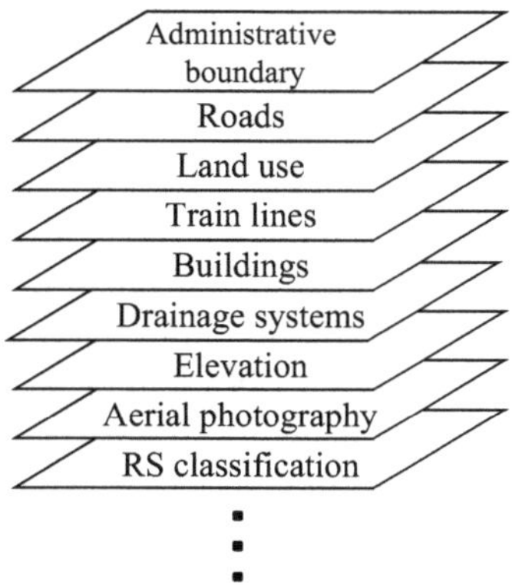

Fig. 1.5 Conceptual outline of GIS layers

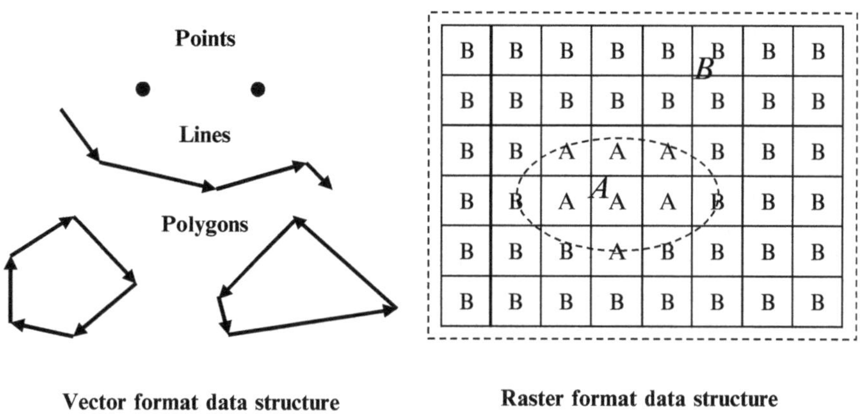

Fig. 1.6 Conceptual diagrams of GIS data structuring

increases the amount of data used. The data structure that one chooses depends on the purpose for which it is to be used. For example, if one is handling polygonal surface data such as buildings, rice paddies, or forests and associated attribute data, one will likely use the vector data structure. Information related to other factors, such as elevation, is in the form of raster format data. The maps, aerial photography, and remote sensing classification results that may serve as the background for GIS displays take the form of raster format data, which also include RGB color information as part of the image data. At present, the points, lines, and polygons of the vector format data structure, which is most commonly used in GIS, are managed as relational databases. Considering that lines are what connect one point with another and polygons are what those lines encompass, these databases are constructed by relating data about points, lines, and polygons. To take full advantage of the characteristics of both vector and raster format data, many GIS products are compatible with both data structures.

1.1.5 Acquisition of Positioning Information

To handle space information in GIS, positioning information is needed. For example, when establishing a plot (investigation division) for field surveys, the shape of the plot is measured from points of reference called benchmarks. However, to use these plots in GIS, accurate positioning information about the benchmarks is needed. Until recently, such positioning information was calculated on maps using a compass or measured from clearly established locations (such as bridges, crossroads, benchmarks) on maps. In addition, it was difficult to obtain positioning information due to the sheer number of places in which photographs had been taken or information gathered. However, with satellite positioning systems such as the Global Positioning System (GPS) now becoming more available, positioning information has become much easier to obtain.

There are two types of satellite positioning systems: GPS, which is managed by the United States Department of Defense, and the Global Navigation Satellite System (GLONASS), which is managed by Russia. These satellite positioning systems were developed for military use in the 1960s and have since come into common use as one of the many instances of civilian use of military technology. Civilian use of GPS began in 1993 as a navigational aid for shipping. In 2000, scrambling "noise," which had been intentionally inserted into GPS signals to maintain the superiority of military systems, was lifted, greatly improving the accuracy of positioning and allowing for progress in the civilian usage of GPS.

The positioning methods used in GPS, which is the most widely used satellite positioning system in the world, are shown in Table 1.1. These methods are generally divided, based on the number of receivers used for positioning, into the following two categories: the independent positioning method (or single-point positioning) and the differential positioning method. Differential positioning is further divided into two methods: one which uses the distance between a satellite and an observation point, which is called "false distance," and one which uses the frequency phases of the carrier (radio) waves. The positioning method to be used is selected according to the purpose and desired accuracy of observation.

The independent positioning method uses only one receiver for positioning and is most commonly used for navigation or simple surveying. This method seeks the position of the point of observation, which is an unknown value, by measuring the false distance between that point and the positions of GPS satellites, which are known values. The distance between a satellite and the point of observation is calculated based on the time differential between the time of transmission of a signal from a satellite and the time of the signal's reception by a receiver. Because the accuracy of the receiver's internal clock is far below that of a GPS satellite, the margin of error of the receiver's clock becomes an unknown value, which, along with the sought-after spatial position (in three dimensions) of the observation point, brings the total number of unknown values to four. Thus, four GPS satellites are necessary for positioning. Although this method can position in real time, the measurement accuracy is comparatively low, with a margin of error of about 10 m.

Table 1.1 Methods of GPS positioning

Type		Name	Accuracy	Field of application
Independent positioning		Single point positioning	10 m	Basic positioning, basic surveying
Differential positioning	Translocation method (using false distance)	Differential method	1 m	Navigational aid, basic surveying
	Carrier phase positioning (using carrier wave phase)	Static method (Static surveying)	5 mm	Benchmark surveying
		Condensed static method	1 cm	Benchmark surveying
		Kinematic method	2 cm	Topographic/ Public works surveying
		Real-time kinematic method	2 cm	Topographic/ Public works surveying

The differential positioning method uses multiple GPS receivers simultaneously. One receiver is placed as a benchmark in a fixed position at known coordinates. Another GPS receiver, which is set at the point of observation, measures signals from the same satellite at the same time as the benchmark receiver. This approach eliminates the primary cause of error, which is shared by both receivers, thus making this a highly accurate method of positioning.

In differential GPS, positioning information is measured at a base point with a clearly established position and the margin of error involved in the observation is calculated. The positioning information observed by a GPS receiver at an unknown position is then corrected using the error information from a receiver at a known positioning point, making it possible to increase the measurement accuracy to within several meters. This method is used for positioning ships in shallow waters and harbors.

Carrier phase positioning is a method which measures the wave phases of the carrier (radio) waves that are transmitted by GPS satellites and the baseline vector between the known position of a benchmark and a measurement point. In the static positioning (static surveying) method, data are collected and analyzed using static antennae affixed at measurement points. These measurements are highly accurate, and thus this method is often used for scientific research activities such as measuring the movement of the earth's crust. Although the accuracy of positioning using kinematic measurement is inferior to that of static surveying, one advantage of kinematic surveying is the ability to make observations of multiple locations in a relatively short amount of time. Therefore, this method is commonly used for surveys of topography or large-scale public works.

Because the earth is not actually spherical, different projection systems called map datums have been developed to assign spatial data to their correct position on the globe. These map datums are specific to regions of the world. In Japan, the

Tokyo Datum geographic coordinate system, which had been in use since the Meiji Period (1868–1912), was revised in 2002 to incorporate the International Terrestrial Reference Frame (ITRF), becoming the "Japanese Geodetic Datum 2000." Two factors set the background for this revision: the development of GPS and other methods of satellite surveying which allow for the acquisition of both precise and readily available spatial coordinates and the increasing need, as evidenced by the expanding use of commercial airplanes or by GIS applications, for a universally shared geographical coordinate system to standardize spatial coordinate data, which had previously varied widely from country to country. GPS uses a geographical coordinate system called the World Geodetic System 1984 (WGS84), which is virtually identical to the current ITRF reference system. In addition, the types of positioning information sought in GPS positioning are latitude, longitude, and "height." However, "height" is different from elevation and thus requires conversion. In Japan, such a conversion program is also available. Also, when verifying positioning using GIS or using GPS in conjunction with maps, it is necessary to determine whether the maps being used employ the ITRF or WGS84 geographical coordinate system. The margin of error between these new maps and the older Tokyo Datum map is a few hundreds of meters.

The methods for obtaining information using GPS are generally divided into two categories. The first method is to obtain positioning information directly by reading the indicator display of a GPS receiver. This method is useful for verifying one's location in the field or for obtaining positioning data. The other method is to record GPS positioning information continuously. This method is used to analyze the movement and behavior of bodies in motion, such as animals or vehicles. GPS information (National Marine Electronics Association, NMEA data) from a GPS receiver is output directly to information recording equipment. For analysis of a body in motion, NMEA data are recorded at fixed time intervals and these data are then analyzed. Such devices are called GPS data loggers. The use of such data loggers to analyze the behavior of living things will be covered in greater depth in Chap. 2. In addition, if one wants to obtain positioning data for photographs, one can very easily obtain such information in post-processing if the internal clock of a digital camera and that of a GPS data logger have been synchronized.

1.1.6 Processing Space–Time Data

The processing of space–time data can be roughly divided into the following five steps: data collection, preprocessing and data input, database management, information analysis processing, and information output. "Data collection" is the collection of spatial information in relation to the target and research purpose. In many cases, existing and relevant resources, such as maps, statistical information, field reports, and aerial or satellite photography of a target area, can be collected. Recently, a wealth of information has become available on the Internet, making a simple Internet search an effective means of collecting information prior to conducting field work. One may also need to collect data independently in accordance with a research theme, and thus sampling or census surveys may need to be conducted.

"Preprocessing and data input" is the task of inputting collected data into a GIS. The digitalization of maps and other data, formatting, error detection and correction, geometric correction, position adjustment, data supplementation, and other such processings are necessary. Although digital data, such as digital maps, are increasingly available, land registries or similar analog data are still quite common and require the greatest expenditures of time and effort to handle in a GIS. Data input and revision are major issues relating to the normalization and continuation of GIS.

"Database management" is an aspect of processing that depends on the systems being used. The fundamental elements of a database management system include controlling data values and definitions and ensuring the preservation, security, and synchronization of data. For the average GIS user, this system serves as a kind of "black box," meaning that the user does not usually interact directly with the system itself. However, these systems provide high-level operational functions and form a critical backbone to support the GIS usage environment.

In "information analysis processing," various types of data are analyzed and processed using a database constructed for that purpose. There are many methods for analysis, such as processing multiple overlaid layers to create new layers, re-encoding attributes, spatial integration, extracting a surrounding area called a buffer, network analysis, and various types of statistical analysis. As illustrated in Fig. 1.7, in overlay processing, layers with different themes are overlaid to create composite layers. This process is not simply overlaying; it also allows one to define new categories based on the combination of two or more different categories, thus creating entirely new thematic layers. As illustrated in Fig. 1.8, buffering is a process of extracting target objects within a set radius from a specific location (point), road (line), or other feature. In addition to these analysis methods available in GIS software, users may also write their own programs to perform analysis. The data resulting from this analysis processing can be added or output as new layers and have many applications as a new source of information.

Generally speaking, the true value of GIS is in the expression of these methods of data processing, to the extent that GIS itself can be thought of as nothing less than "data analysis processing."

"Information output" is the operation in which the final analysis results are output. GIS can output many kinds of maps and diagrams. Outputs include thematic maps which illustrate across-the-board features, choropleth maps which illustrate the relative degree of some continuous variable, and contour maps and various statistical diagrams. In addition, analysis results can also be output to files and network-based transmission for use in analysis with other systems.

GIS is an extremely valuable tool for managing space–time information. However, because of the high costs of constructing and maintaining these systems, and also the high degree of expertise required for their use, at present these systems are not very widely used outside specialized fields. However, GIS can easily be used with a common personal computer. A wide variety of commercial GIS software is

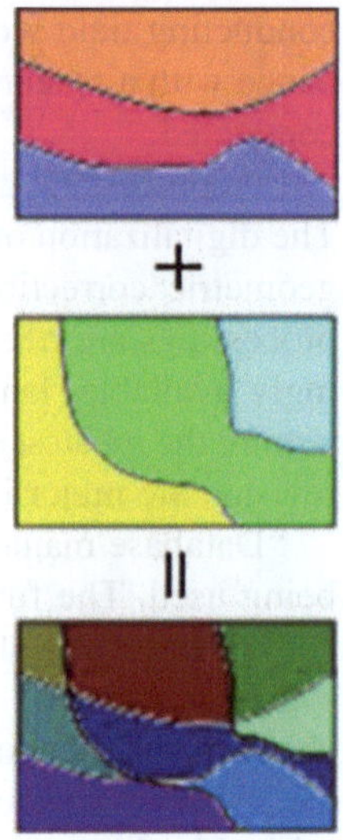

Fig. 1.7 Overlaying

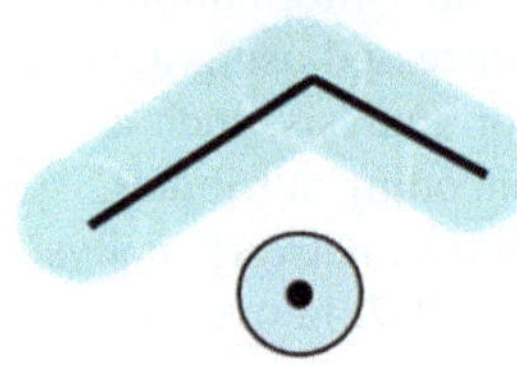

Fig. 1.8 Buffers (*shading*)

also available, ranging from software systems with high-level functionality to those which are simple and easy to use. Furthermore, free, open-source software is also available. In addition, in terms of map data for GIS, Digital National Land Information and digital maps produced by the Geographical Survey Institute and other such organizations are available for use. Thus, progress is being made in laying the groundwork for GIS implementation, with high expectations for more widespread use of GIS in the future.

The Internet is a rich source for various types of information, such as map data and travel route data. Promotion of digital data was sparked in 1998, when Al Gore, then the Vice President of the United States, proposed the inception of what he called "Digital Earth." This is the notion of using information technology to relate the actual earth to a simulated earth in an attempt to virtually reproduce, and thus better understand, our planet. Its originally intended applications were in fields such as education, research, land use, urban planning, and environmental conservation. To construct and implement such a system, technical advances such as high-speed networking, high-resolution satellite photography, and the acquisition of accurate positioning data through GPS were all necessary. Today, "Digital Earth" has been realized, and its refinement continues. Digital Earth sites on the Internet include "Google Earth" (Google), "Virtual Earth" (Microsoft), and "World Wind" (NASA). Map search engines, such as "Google Maps" and "Mapion," are available as well.

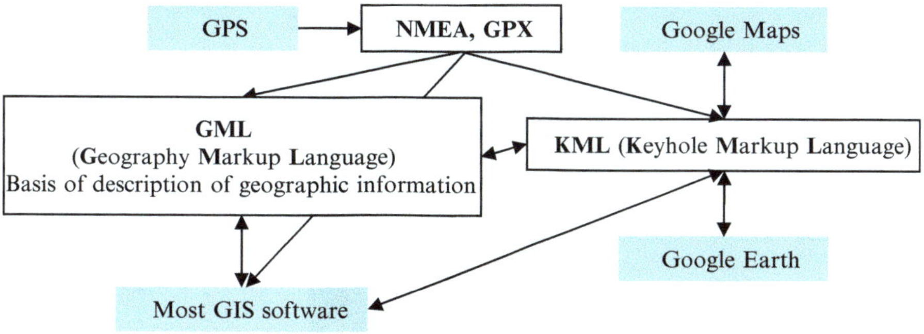

Fig. 1.9 The standardization of and interrelation between spatial data formats

The standardization of data formats for data created by various types of GIS software provided the background for the spread of such Internet-based map data. These formats have been standardized by the International Organization for Standardization (ISO) as ISO19136 (GML3.1). This type of data format standardization is opening up new uses for spatial information. Figure 1.9 illustrates the relationships between the data formats used in GPS, GIS, and similar systems, with arrows representing the flow of information. Most GIS software is compatible with the Geography Markup Language (GML) format as well as with the NMEA format and GPS Exchange Format (GPX), which are the data output formats used in GPS. On the other hand, "Google Earth" and "Google Maps" use a data format called Keyhole Markup Language (KML) to describe the points, lines, polygons, and images that make up space information. The Open Geospatial Consortium, which aims to establish worldwide standards and worldwide interoperation of spatial data, designated KML as their standard for spatial data. While there are some differences between the GML and KML data formats, they are highly compatible.

It is very important to integrate, using GIS, the results obtained through the analysis of remote sensing or field surveying, and to make these results available to the public or at least to interested parties. One means of doing so is to make data available on the Internet. Figure 1.10 illustrates an example of space information managed by GIS which has been transferred onto the Internet-based "Digital Earth." This example uses "Google Earth" and "Google Maps" as its "Digital Earth." The point, line, and polygon vector data and the corresponding attribute data, created using GIS, have been output as KML format data and fed into "Google Earth" and other sites. Figure 1.10a shows a screenshot of a GIS display, Fig. 1.10b shows the screenshot after being fed into "Google Earth," and Fig. 1.10c shows the screenshot after being fed into "Google Maps." In this way, it is possible to provide research information about natural resources or the environment on the Internet. The Internet thus offers a way to transmit spatial information and possibilities for the sharing of, and applications for, information about natural resources and the environment covering the whole earth.

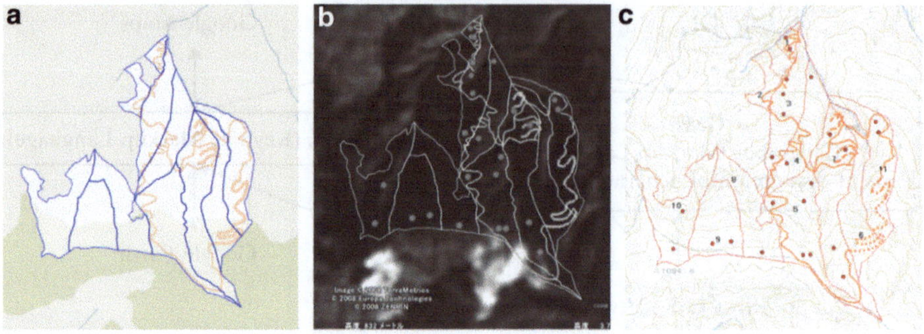

Fig. 1.10 Example of image transfer from a GIS image (**a**) to Google Earth (**b**) and to Google Maps (**c**)

1.2 Case Studies of Remote Sensing and Geographic Information System Usage

To understand the current state of the natural environment on Ishigaki Island and Iriomote Island, Japan, land cover (land use) was analyzed using remote sensing. Remote sensing analysis begins with the acquisition of satellite image data. Here, imagery from the ASTER sensor of the Terra satellite was used. The ASTER-VNIR sensor has a resolution of 15 m and three bands: visible (green and red) and near infrared light. Figure 1.11 shows the raw image and the same image after geometric correction processing. The island in the upper left of the figure is Iriomote Island and that in the upper right is part of Ishigaki Island. This picture was rendered using false color, with the near infrared band assigned the color red, the red band assigned the color green, and the green band assigned the color blue. Thus, on the basis of reflection characteristics, the red areas are vegetation, the whitish areas are urban areas or bare fields, the blue areas are shallow seas, and the green area is the open sea. The image created by this sensor has positional information attached at its four corners and center, and Fig. 1.11b displays the result of geometric correction. The image has been revised so that it matches the geographical coordinate system of the map.

Next, the target area for analysis was selected from the geometrically corrected image and unsupervised classification was performed. Figure 1.12 shows a false-color image of the target area (a) and the classification result for land cover (b). The variables used were the green band, red band, near-infrared band, and the NDVI vegetation index. In Fig. 1.12a, the NDVI is assigned the color red, the near-infrared band is assigned the color green, and the red band is assigned the color blue. Thus, highly vegetated areas, in which the NDVI and near-infrared band values are high, are displayed in yellow. The classification result displayed in Fig. 1.12b classifies Iriomote Island into forests (shades of green), the coral reef around the island (light blue), and urban and farm areas (pink). Using such images spanning different time periods, one could research changes in land use and land cover on the island.

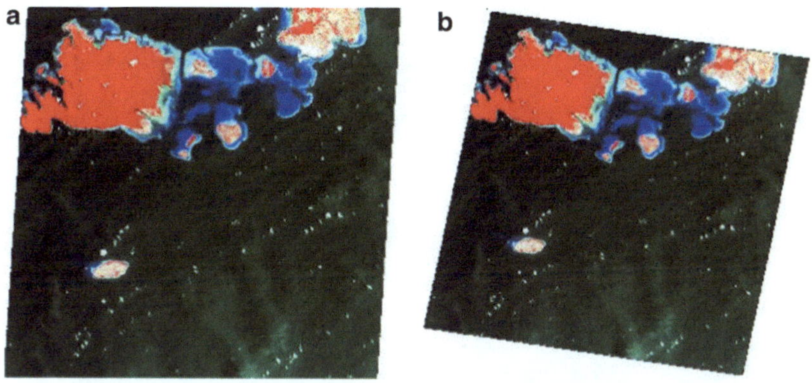

Fig. 1.11 Raw image (a) and image after geometric correction (b)

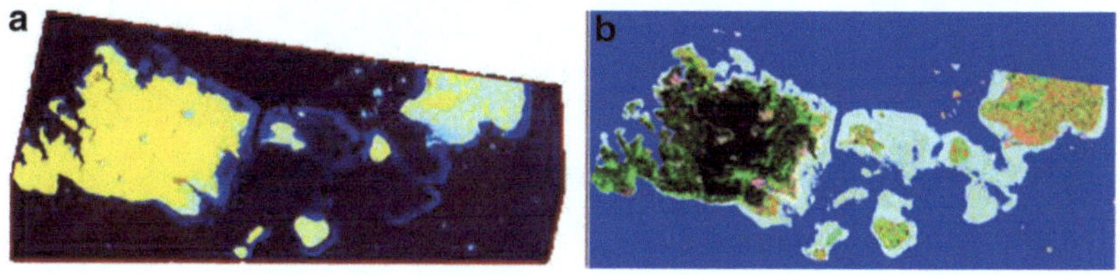

Fig. 1.12 False-color image (a) and image after unsupervised classification (b)

Next, the differences in land cover between two drainage basins on Ishigaki Island were considered. Figure 1.13 is the land cover classification result of Fig. 1.12b as handled using GIS. The classification map for land cover is displayed transparently, overlaid onto the corresponding map from the Geographical Survey Institute, in the light-purple square at the center of the image. Also, the two drainage basins in which land-cover differences will be considered are displayed as polygonal data at the center of the image. In GIS, space information is managed at the layer unit level. In this example, three different layers are being used: the Geographical Survey Institute map (raster format data), the remote sensing classification result (raster format data), and the drainage basin area polygons (vector format data). Also, in a GIS, the user can select at will which layers to display or not display.

The Geographical Survey Institute map and the drainage basin polygons are indicated as they appear in the GIS in Fig. 1.14. Using this polygonal information, the two pictures on the right side of Fig. 1.14 display the land-cover classification results for each drainage basin. The differences in coloring, which indicate different classification classes, show that land usage differs between the two adjacent drainage basins, A and B. Because each pixel in a satellite image can be converted to a certain fixed area, it is possible to calculate the area for each type of land use by counting the number of pixels. These results can serve as fundamental information for understanding, for example, river turbidity and its impacts, based on differences in land cover.

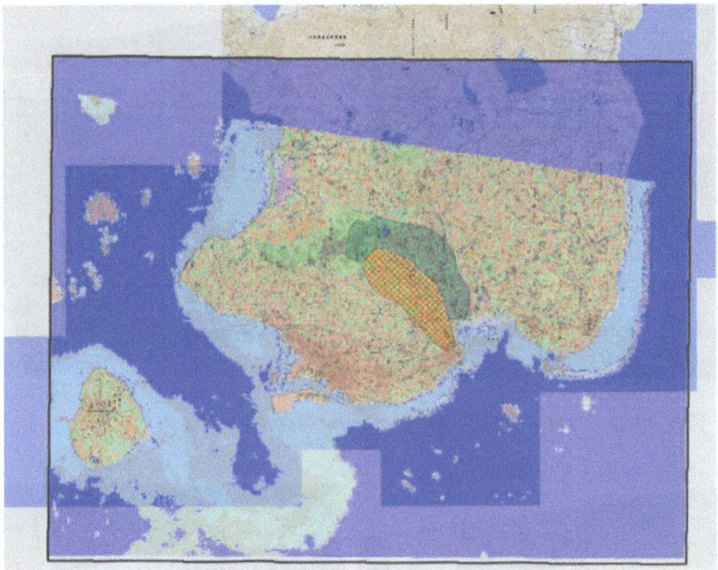

Fig. 1.13 Land cover classification in GIS

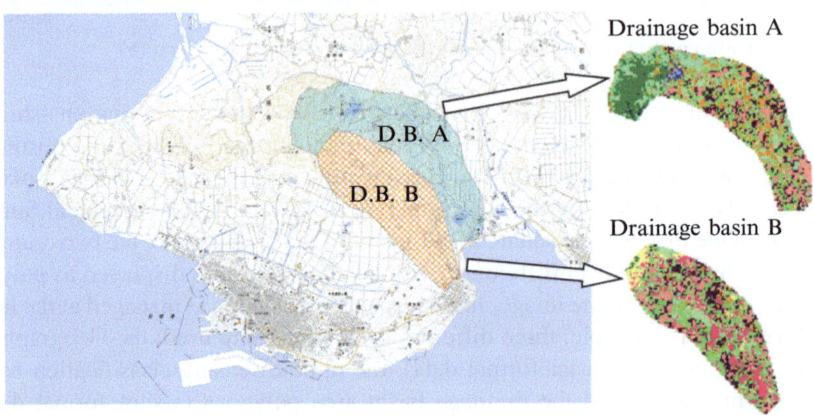

Fig. 1.14 Extraction of the land-cover classification results for the two drainage basin areas

GPS is an invaluable tool for obtaining location data. By utilizing multiple GPS receivers or GPS data loggers, it is possible to obtain location data continuously. These data can be used in a GIS as route data or location data. Figure 1.15 shows an example in which the routes traveled by mountain climbers were recorded using GPS for examination of the distribution of visited routes and travel frequency. At the two trailheads of a mountain (distinguishable as red and blue lines), visiting mountain climbers were asked to carry GPS data loggers, and information about the routes that they traveled was gathered. The deeper-colored lines indicate areas

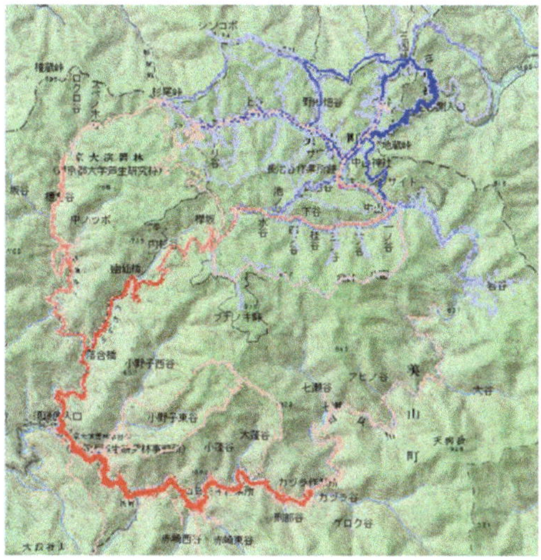

Fig. 1.15 Routes traveled by mountain climbers (Original image by Yoshimura Tetsuhiko)

through which many mountain climbers passed, providing fundamental data for devising strategies to protect the natural environment or regulate access to the mountain. By identifying which areas mountain climbers were visiting, these data also proved useful in considering where guideposts and other installments were needed. Using these types of methods, GPS and GIS are used to analyze the behavior of bodies in motion. Recently, these techniques have also been applied to the movements of wild animals, such as bears, deer, and monkeys, and are becoming an important source of information for designing measures aimed at allowing humans and wild animals to live in harmony.

Remote *s*ensing, geographical information *s*ystems, and *s*atellite positioning systems are referred to as the "three S technologies." The collection and analysis of spatial information and the subsequent results are useful "tools" in real world applications. Information systems that make use of spatial information range from global-level, wide-scale systems such as those used for analyzing climate change, to regional-level, small-scale systems such as those used for disaster prediction and prevention and for environmental mapping. Spatial information science bringing real-world applications to geographic information and the "three S technologies" are expected to continue to grow in importance in the foreseeable future. Please refer to the References Section for more detailed information about remote sensing image processing (Japanese Remote Sensing Research Group 1981; Takagi and Shimoda 2004) and for more detailed examples of usage (Kato 2007; Akiyama et al. 2007).

References

Akiyama, T., Ishitsuka, N., Ogawa, S., Okamoto, M., Saito, G., Uchida, S. *Agriculture Remote Sensing Handbook* (in Japanese). Japanese Agricultural Systems Society, 2007

Japanese Remote Sensing Research Group. *Processing and Analysis of Images* (in Japanese). Kyoritsu Publication, 1981

Kato, M. *Forest Remote Sensing* (in Japanese). Forestry Investigation Meeting, Tokyo, Japan, 2007

MultiSpec: http://dynamo.ecm.purdue.edu/-biehl/MultiSpec Japanese manual (http://www.affrc.go.jp/rss/2004/colorPPT.pdf)

Takagi, M., Shimoda, H. *Image Analysis Handbook* (in Japanese). Tokyo University publication, Tokyo, Japan, 2004

Chapter 2
Biologging

Nobuaki Arai and Junichi Okuyama

Abstract In recent years, the habitats of wild animals and humans have overlapped remarkably as economic activity intensifies and expands. As a result, human activities in the biosphere have brought about a decrease in biodiversity and have become a destabilizing factor of the ecosystem; these can no longer be ignored. Human activities cause various frictions in the ecosystem, which in turn cause serious social problems such as damage to residents, agriculture, and fisheries. This is a pressing issue to be solved if society is to achieve sustainable use of natural resources. Thus we obliged to assess human impacts on the ecosystem and reclaim our role as a member of the biosphere in the ecosystem. Towards this cause, using behavioral information about wild animals helps to clarify their relationship with man. In this chapter, we introduce methods for acquiring animal behavioral information using the latest information technology.

2.1 Acquisition of Animal Behavioral Information

Visual observation is a basic technique to survey wild animal behavior. However, direct observation of wildlife is generally difficult because of their high mobility, nocturnal activities, and the dangers of accessing their habitat, among others. In addition, it is impossible to observe them in the hydrosphere, which accounts for about 71% of the earth's surface, including lakes, marshes, and rivers. Therefore, the collection of behavior information on wild animals living in these areas has been overlooked for the simple reason that direct observation is difficult.

Recently, researchers have begun to observe wild animals using a Cessna or helicopter, which makes large-scale observation possible. This is effective for

N. Arai (✉) • J. Okuyama
Graduate School of Informatics, Kyoto University, Kyoto 606-8501, Japan
e-mail: arai@i.kyoto-u.ac.jp; okuyama@bre.soc.i.kyoto-u.ac.jp

T. Ishida (ed.), *Field Informatics*, DOI 10.1007/978-3-642-29006-0_2,
© Springer-Verlag Berlin Heidelberg 2012

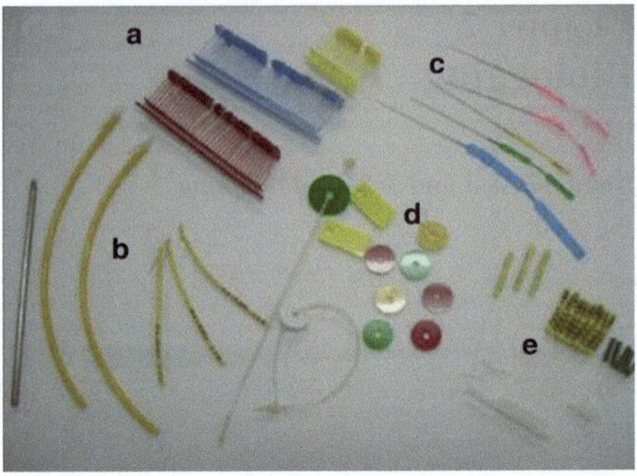

Fig. 2.1 Various tags are used to measure the movement of aquatic animals: (a) anchor tags, (b) dart tags, (c) ribbon tags, (d) Atkins tags, (e) spaghetti tags

observation of aquatic animals. The realities of the biological community, which have far exceeded our imagination, have been illuminated through observation even in the deep ocean by submersibles or remotely operated vehicles (ROV). However, because such observation cannot be conducted over long periods, we are able to get only a snapshot of animal activities.

As for follow-up observation, tags have been attached to target animals to understand how they move from the release point to the recapture point. For example, the pathways of migratory birds are investigated using tags to assess the effects of sea farming on their behavior. This brings us the basic information on their habitats and migration route. A variety of tags are used according to species and size (Fig. 2.1).

We get only the location information at release and recapture point in a tagging survey. On the other hand, wild animals move around in their habitats throughout their life histories. If environmental information that they experience can be measured in detail, it becomes possible to guess their behavior based on this background. The desire for this type of information is one of the driving forces in the development in biologging technology.

2.2 Techniques of Biologging

Biologging is a technique to observe animals in their habitat where we have limited access. Japanese scientists and engineers have been playing an advanced role in the technological development of biologging. Roughly, there are two methods of biologging. One is a method to accumulate information with small data recorders (micro data loggers or archival tags) attached to or implanted in the animals. Archived data can be obtained after collecting the recorder. The other is a method of installing a beeper and an ultrasonic transmitter in the animals, which is called biotelemetry. It is possible to acquire data on the movement and physiology of object animals, in addition to the environment they experience. Prof. Naito of the National Institute of Polar Research and Prof. Sakamoto of Kyoto University developed the data logger in the 1980s, and, they succeeded in recording the behavior of sea turtles and other marine animals. A digital data logger miniaturized and can be installed in a number of animal species. As for the latter, biotelemetry with a small beeper has been used for birds with a vast home range since 1970s. Biotelemetry with an ultrasonic transmitter has been used to measure fish behavior in water.

The word biologging was first used at an international symposium held at the National Institute of Polar Research in 2003 (Naito 2004). The Japanese Society of Biologging Science was organized in 2004, and since then biologging has been an established technique for behavioral analysis of animals. In recent years, we are able to obtain a large amount of data through data loggers, thanks to multi-itemization of the sensor and increase of the memory capacity. Information technology makes a significant contribution to improvements in data analysis.

From the viewpoint of observing action and activity, biologging can be understood as a kind of human-sensing information technology. In human-sensing, the object is the same being as the observer. The observation site can be artificially set. It is easy to understand acquired data because the observer can communicate with the testee. On the other hand, biologging observes the behavior of animals living in places where it is impossible to access. The measuring instruments must be able to endure various environments. Furthermore, the installation and settings of the instruments for data acquisition are very difficult. Acquired data sometimes exceeds our expectations, and requires consideration and examination from all aspects.

2.2.1 *Acquisition of Behavioral Information Using Micro Data Logger (Archival Tag)*

The development of micro data loggers (archival tag) began as a clockwork-powered analog process writing data to a recording chart stored in a waterproof applicator. It was reduced in size to below 10 cm in total length in the 1980s. Analog data loggers were first attached to penguins, seals and sea turtles with easy collection in the late 1970s. In addition to the diving behavior of subject animals, other data such as a water temperature was obtained. In the 1990s, the recording medium changed from paper to electronic memory, an epoch-making increase in storage capacity, and various sensors were loaded. This made it possible to acquire data that needed a more high-speed sampling rate, such as acceleration and terrestrial magnetism (Table 2.1). This is commonly called micro data logger or archival tag. At first, it was called Time-Depth Recorder (TDR) since sensors were commonly used for only acquiring water temperature or hydraulic pressure data. There are no problems in practical use in data sampling with recording intervals of about one time per second for water temperature or depth. However, there are restrictions imposed by the capacity of built-in memory and battery life. Memory capacity and

Table 2.1 Sensors installed in a micro data logger and obtainable information

Sensor	Obtainable information
Pressure	Depth
Temperature	Ambient/body temperature
Acceleration	Movement, posture
Propeller	Convection velocity
Light	Light intensity, rough location
Magnetic	Geomagnetism, azimuth direction
Electric potential	Heart rate, myogenic potential, brain wave
Image	Still image, moving image
GPS	Location

battery life still has a problem, even now, because the micro data loggers are used for the purpose of obtaining image and sound information. The latest micro data loggers used at present are listed in Fig. 2.2.

The micro data loggers can obtain the data of physiology activity of subject animals, and environmental information as well. Recently, GPS data loggers using Global Positioning System (GPS) were developed, and the location information of object animals can be acquired. Because the signal from a GPS satellite cannot be received in water, the GPS data logger cannot be used for aquatic animals except for animals surfacing when they breathe, like sea turtles and marine mammals.

All data is stored in built-in memory in the micro data logger. Unless the data logger is collected, data cannot be acquired. In general, it is very difficult work to collect the released animals again. Therefore, micro data loggers have been attached to nesting birds and sea turtles expected to return to the same place. Automatic devices that release the data logger after a fixed time (Fig. 2.3) were also developed. The micro data logger was also collected by the telemetry technique. In this way, development of biologging techniques has expanded (Rutz and Hays 2009; Cooke et al. 2004).

2.2.2 Acquisition of Behavioral Information Using Transmitter (Biotelemetry)

Using the micro data logger in the preceding section, we can analyze in detail the behavior of subject animals corresponding to the native habitat and the environment, based on the large and detailed data from the recaptured logger. However, precise position data cannot be obtained without the use of a GPS data logger. Biotelemetry is a technique to acquire the behavioral information with a transmitter, which is useful for measuring the position of the animals in detail. In biotelemetry, the researcher pursues the animal in order to receive the signal from radio waves or the ultrasonic transmitter installed in the object animals. Radio wave transmitters are used on land and in freshwater environments. Because radio waves cannot be transmitted in seawater, ultrasonic waves are used for the observation of fish. Satellite transmitters are used for observation of animals that move terrestrially.

Fig. 2.2 Catalog of the latest micro data loggers (archival tags) (Photos and drawings by courtesy of Little Leonardo Ltd.)

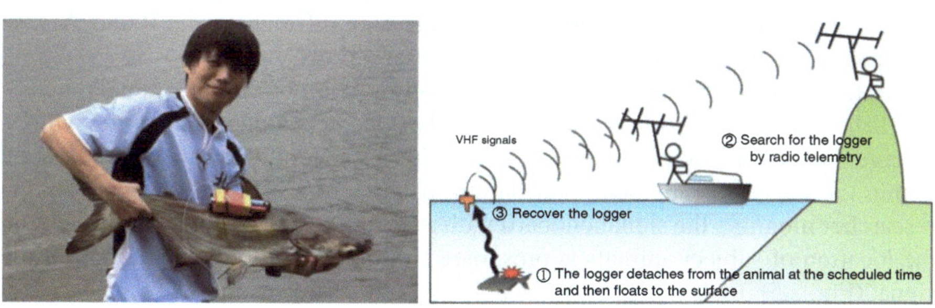

Fig. 2.3 Mekong giant catfish with attached micro data logger and a timer type logger collection device (*left*). Procedure of collection of the micro data logger using an automatic timer type logger collection device (*right*)

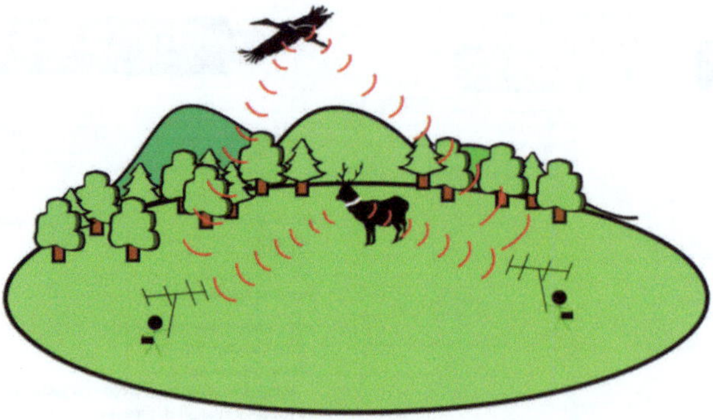

Fig. 2.4 Conceptual diagram of radio wave biotelemetry. Using radio wave telemetry, the researcher receives radio waves from two or more points and locates the sending source (subject animal)

Radio wave biotelemetry is used for terrestrial animals like birds and mammals. The frequency of the radio wave ranges 30–300 MHz, and VHF is often used. A directional antenna such as the Yagi Antenna is used for detecting radio waves. The position of the subject animals is estimated by maximum azimuth of the radio wave from two or more points (Fig. 2.4). An accurate map with appropriate scale is needed to locate them. It draws a straight line to the azimuth of the maximum receiving sensitivity from the position of two or more points. The estimated positions are filled in on the map, and the intersections of those points are the positions of subject animal. When the intersection is not corresponding by three or more points, the polygonal center of gravity can be done using the straight lines. When researchers use biotelemetry, they carry the receiver and the antenna, and pursue the signal source. If the antenna and the installation type receiver can be fixed, allowing animals to be tracked with the range of reception of the base station (Cooke et al. 2004). Recently, a multitelemetry system has been developed. Two or more base stations are set up and connect them by a communication network. It makes it possible to clarify the position of the animals within the range of reception with high accuracy. However, it should be noted that some of the units may infringe on the Radio Act, so it is necessary to pay attention in making observations.

Biotelemetry with ultrasound is used for observation of marine animals including fishes. There are two tracking methods. One technique is tracking with a receiver on board a ship. Another is a technique tracking with receivers arranged within the expected route and distribution of the animals. In the former case, the researcher monitors the signal onboard with suspending a hydrophone in water, and the location of subject animals is presumed from the direction and strength of the signal (Fig. 2.5). In the latter case, the presence of the subject animals can be known by downloading the data accumulated in a built-in memory of the fixed-type receiver (Fig. 2.5). The ultrasonic transmitter currently used with unique identification codes (ID) is adjusted by the interval time of ultrasonic wave pulse (69 kHz).

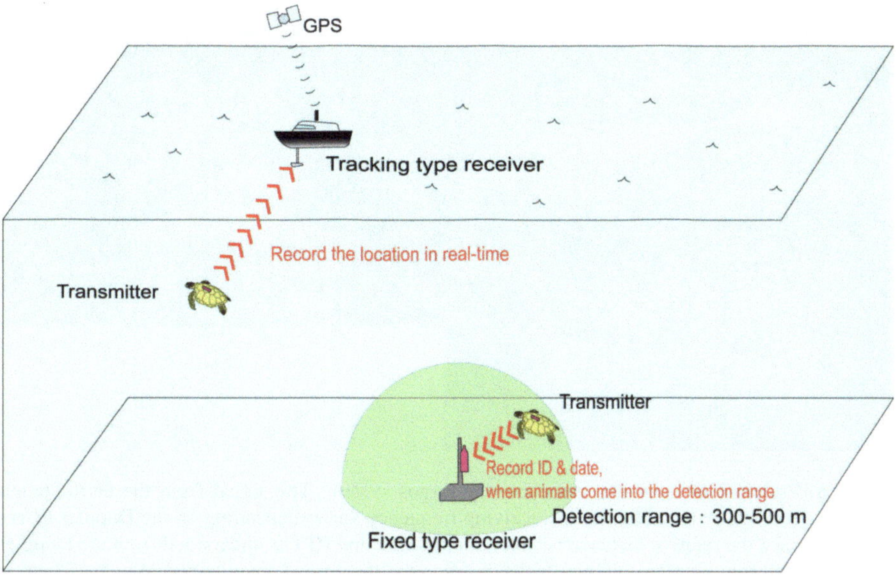

Fig. 2.5 Conceptual diagram of ultrasonic biotelemetry. The subject animals are tracked in real time using the tracking receiver onboard a ship. On the other hand, fixed receiver records transmitter ID and the detection time of the animals within the range of the reception

This can pursue several individuals at the same time even with ultrasonic wave signals of the same wavelength. It is also possible to measure the water temperature or depth by setting the ultrasonic wave pulse interval value.

A real-time positioning system that can pursue the behavior of the animals within the reception range with high accuracy by recording the ultrasonic wave signal at the same time with the installation type receiver at three or more points has been developed.

It is extremely difficult to pursue animals with a broad range of movement such as birds, whales or sea turtles, with radio or ultrasonic telemetry. These animals can be pursued with the Argos system using space satellites. The Argos system was developed by the French Centre National d'Études Spatiales (CNES), the United States National Oceanic and Atmospheric Administration (NOAA), and the National Aeronautics and Space Administration (NASA) in the 1970s. It is maintained and managed with the cooperation of France, the United States and the Japan Aerospace Exploration Agency (JAXA). The principles of measurement by the Argos system are summarized as follows. The space satellite orbits the earth every 102 min at an altitude of about 850 km. It receives the signal (401.650 MHz $\pm$ 4 kHz) of the Platform Transmitter Terminal (PTT) from about a 5,000 km diameter on the ground. The signal that the satellite receives changes by the Doppler Effect by the relative speed of the satellite and PTT. The distance between a certain momentary satellite and PTT is obtained to take into account the

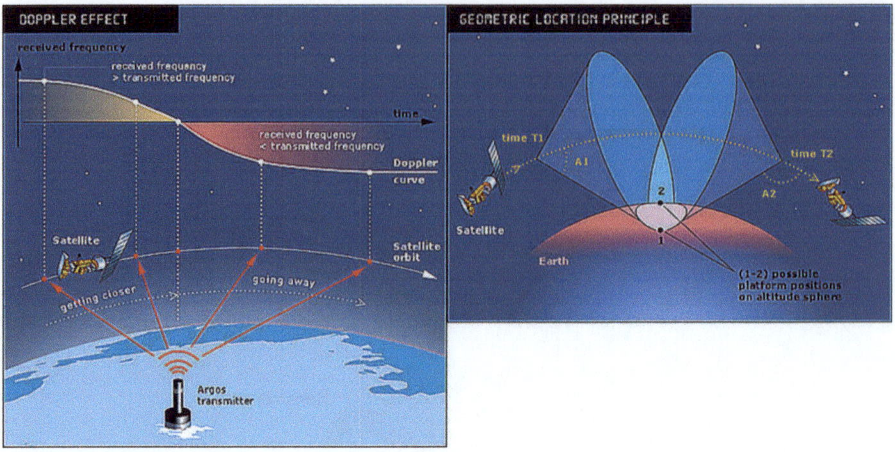

Fig. 2.6 Principle of measurement using the Argos system. The signal from the earth station (PTT) is received by the satellite. Its receiving frequency varies depending on the Doppler Effect (*left*). Because the relative distance between the satellite and PTT is understood from the Doppler Effect, the intersections (two points) of the distance and the ground are obtained. Which of these is appropriate between two points is judged from the discharge point (*right*)

frequency variation of the signal by this Doppler Effect. The intersection of the distance and the ground becomes the position of PTT (Fig. 2.6). Thus, PTT on the ground only sends the signal including ID to the satellite, and the Data Collection and Location System (DCLS) installed in the satellite calculates the position. After the downlink to the earth station (about 50 stations worldwide), the location information is sent to the user.

The Argos system can obtain location information everywhere because the data is transmitted to the user at the same time as measuring the position. The location information is divided into classes (3, 2, 1, 0, A, B, Z) depending on the accuracy. The highest accuracy class 3 is guaranteed for an accuracy of less than 150 m, although the positional precision is inferior to GPS. Accuracy worsens greatly in class A or B, and whether to use the positional data is entrusted to the researcher's judgment. Therefore, satellite tracking using the Argos system is used for the observation of animals with broad range migrants without good accuracy.

The new tracking system combining various data loggers with the Argos system is more advanced, and transmits the data obtained by data loggers through the satellite communication system of the Argos system. For instance, the GPS-Argos system, which combines a GPS receiver and an Argos transmitter, was developed to offset the poor accuracy of positional measurements of the Argos system by accurate location of the GPS (Fig. 2.7; Yasuda and Arai 2005). The pop-up tag system that was built by combining the data logger and the automatic surfacing system into the Argos system was also developed. This makes it possible to clarify the broad range migration of large fishes that live outside the range of radio waves.

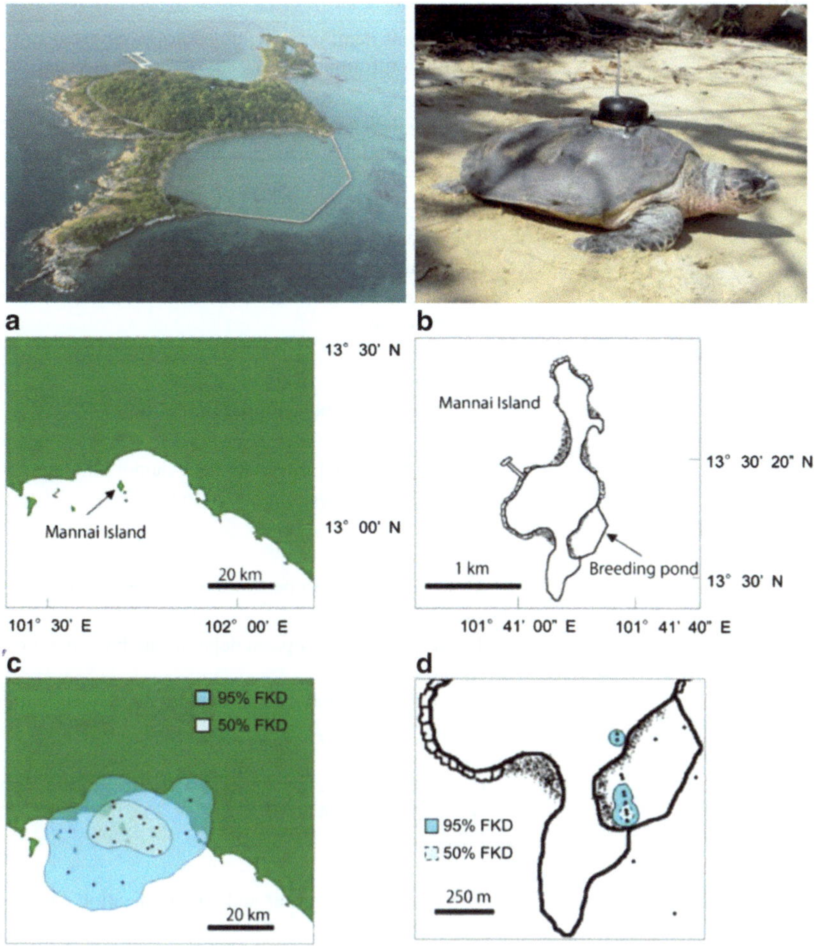

Fig. 2.7 Experimental tracking of hawksbill sea turtles with GPS-Argos transmitters was performed at the sea turtle protection research facility of the Department of Marine and Coastal Resource at Mannai Island, Thailand (*upper left*). A GPS-Argos transmitter was attached to the carapace of hawksbill turtle (*upper right*) (HB1). (A) and (B) indicate the maps of Mannai Island and its surrounding sea area. Based on the GPS-Argos transmitter results the home range of the turtle almost distributes within the breeding pond (D). To the contrary, however, the Argos system results indicated the home range extends to the surrounding area of the island beyond the breeding pond (C) ((A–D) are reprinted from Yasuda and Arai (2005) with permission)

2.2.3 Features of Biologging Techniques

The techniques that use the micro data logger (archival tag) and the technique of biotelemetry with a transmitter have advantages and disadvantages respectively (Table 2.2). Based on a good understanding of the features of these techniques, it is necessary to select the more appropriate technique when biologging is used.

Table 2.2 Features of each biologging technique

Micro data logger (archival tag)	Measurement item: activity, physiology, environment, and location information
	Merit
	In-depth data can be taken regardless of time or the place
	Labor is minimal as long as the collection method is established
	Demerit
	The data logger is expensive
	If the data logger is not retrieved, data cannot be obtained
Transmitter	Measurement item: location, environment, and activity information
Radio/ultrasonic (tracking type)	Merit
	The transmitter is relatively cheap
	Information can be acquired in real time
	Demerit
	If the signal is not received, data cannot be obtained
	It is labor intensive and costly to track
Radio/ultrasonic (fixed type)	Merit
	Labor is minimal
	Multiple targets can be tracked at the same time
	Demerit
	The range of the reception depends on the number of base receivers
	Data cannot be obtained if there are no subject animals within reception range
Satellite tracking	Merit
	If it is possible to communicate with the satellite, data can be acquired anywhere
	Labor is minimal
	Demerit
	The satellite transmitter and communication charges are very expensive
	The accuracy of the location information is bad

Activity, physiological and environmental information on various animals can be obtained now by techniques that use the micro data logger, as shown in Table 2.1. When terrestrial animals are tracked, detailed location information can be acquired by the GPS data logger. The biggest advantage of this method is to obtain fine-scale location data. Another advantage is that a detailed behavior record can be obtained regardless of place or time. On the other hand, difficulty in recovering the data logger is the worst disadvantage. Therefore, the researcher should make an in-depth research plan assuring that the data logger can be recovered using a timer type logger collection device or similar. Moreover, the expense of the data logger is also a disadvantage.

The technique with a transmitter (biotelemetry) is chiefly used to obtain the location information of the subject animals. In addition, some kinds of transmitters can transmit the information on the environment, animal physiology and activity like depth and temperature data at the same time. As described above, biotelemetry has two types of tracking to obtain data. One is a tracking type in which the researcher carries the receiver during the tracking. The other one is the installation type, where the researcher set up the receivers. As for the tracking type, the advantages are to obtain the data in real time, and at a reasonable cost for the initial investment for the research. However, the tracking type also has disadvantages, in that it requires a huge amount of work to pursue the target animals, and requires high personal cost for tracking. The researcher is not able to obtain any more data if the target animals are lost. The installation type does not require as much work as the tracking type because it only needs regular data download from the deployed receivers. There is also a receiver into which the automatic data transfer system to the researcher is built. A weak point of the installation type is that the researcher is not able to acquire the data at all if there are no target animals within the reception range. In addition, it is costly to obtain the information on wide-ranging movement because the tracking range depends on the number of the receivers.

As for biotelemetry that uses the Argos system, the features are different from those of the radio and ultrasonic telemetry. The merit of this method is that the researcher is able to get the location data wherever the target animal exists if the transmitter can communicate with the Argos satellite. Data obtained are transmitted to the earth station via the satellite, and delivered to the users by email. Therefore, the researcher does not need to recover the device and consequently the labor is very minimal. The disadvantages are that the communication charge for data transmission is high in addition to the charge of the main body of the transmitter. It might cost millions of yen when tracking exceeds 1 year because the satellite communication charge is calculated on a daily basis. Moreover, the position-fix accuracy of the Argos system is comparatively bad as described in the foregoing paragraph. Therefore, it is not for the observation of the animal that does small-scale movement, but for the animals that migrate across areas.

2.2.4 Method of Installing Measuring Instruments

It is necessary to pay close attention to the installation so as not to influence the behavior or the lives of the animals when the micro data logger and transmitter are installed in the animals. There is no clear basis although the size of the measuring instrument should be roughly within 3–5% of weight of the animal (Cooke et al. 2004). When the measuring instrument is installed on the subject animal, narcotics might be used for the safety of the observers and stress reduction for the subject animals. It is necessary to pay close attention to the use of narcotics, and to give subject animals an appropriate dose. The measuring instruments are installed on the back of birds and mammals in general. Other types of measurements are also used; the band type is used for birds, and the collar type for mammals. There are some installation types, such as an external installation, stomach interior wear, and the inside of the body installation for fishes. When the radio transmitter is installed on the subject animal, there is a possibility that the action (behavior) is restricted by the antenna because the antenna part is long. The transmitter is rarely expelled from surgical region after several hours or a few days when it is surgically installed in a body cavity. It is necessary to select the most suitable method by conducting a preliminary installation experiment, and to take into account the effect on the survival rate of animals after surgical operation. For at least 24 h before the experiment, one should check abnormal behavior of the animals that might be caused by the transmitter attachment. Before releasing animals with the micro data logger and transmitter to the outdoors, it is also important to confirm it works normally.

2.3 Analysis of Animal Behavioral Information

The data of the micro data logger are time-series data taken at a constant sampling interval. Time series analysis is needed to clarify the behavior of subject animals and environmental information. In recent years, animal movement obtained from the acceleration data has been observed. Acceleration data are obtained as numerical values from the acceleration by the movement of the animals based on the gravitational acceleration. The posture of the subject animals is estimated from the element of gravitational acceleration. The periodic movement of animals such as flipper/wing stroke of birds, and tail beat of fish can be identified by spectral analysis. In recent years, various sensors of the micro data logger enabled us to acquire a number of behaviors and physiology of the subject animals and their environmental information simultaneously. Behavioral information, including the acceleration data, enables us to classify the behavioral patterns of animals, and to know what, when, and where they do at specific times. Information processing technology such as the cluster analysis and data mining are necessary to ascertain the behavior of the animals from many kinds of data in large quantities.

The home range of the animals can be presumed from the location information obtained by GPS data logger and various biotelemetry investigations. The home range is defined as a place where a certain individual usually uses to feed and breed. Places used only temporarily are not included in the home range. Animals know where there is food and safe retreats. In general, 95% of animal's positions are included within the home range of this animal. When we analyze the location data of animals, it is necessary to consider the objective of the study, the quality of data, and the condition of data collection. It is also necessary to choose a suitable method from the various home range analyses through comparison with other research. Maximum Convex Polygon (MCP), kernel (estimate), harmonic mean, grid cell, ellipse, and Fourier series, etc. are primarily used for estimating the home ranges of animals. The home range and movement data obtained by these methods can be analyzed in more detail by combining them with a variety of environmental information based on geographic information systems (GIS).

2.4 Case Studies of Biologging

2.4.1 Behavior Analysis of Green Turtles Using Micro Data Loggers

The Similan Islands in Thailand of Southeast Asia are a popular tourist spot known for recreational diving and other marine leisure activities. Therefore, the area was designated a national ocean park to protect its natural environment. It is very important to understand how the marine animals living in this region use the protected zone in order to appropriately manage this kind of area as a sanctuary. Therefore, the behavior and ecological information of the green sea turtles that are the symbolic species in the Similan Islands were obtained by using a micro data logger.

We investigated the green sea turtles on Huyong Island, which is one of the Similan Islands, and a nesting area in the rainy and dry seasons of 2003 and 2004. Sea turtles crawl up to the beach and lay eggs two or more times at an interval of 10–15 days during one nesting season. Therefore, if the micro data logger is installed in laying sea turtles, the researcher can recover the data logger at high rates during the next landing, and obtain the data about what the sea turtle does during the internesting period.

The depth data loggers (Little Leonard Co. Ltd., Japan) were installed in the carapace of three green turtle individuals that had laid eggs in March (dry season), June (shift period), and August (rainy season) by using an epoxy resin adhesive (Fig. 2.8). To obtain detailed behavior information, the sampling time of depth was determined to be 1 s. The object individual was captured after egg laying ended in the landing next time, and the data logger was collected.

Figure 2.9 is the time-series data of the stay depth during the internesting period of individual 1 and individual 3. It is understood from this figure that individuals 1 and 3 use the depth zone of 0–40 m and 0–90 m, respectively. Sea turtles need to swim up to the ocean surface for breathing. The interval of the surfacing can be read from the depth data of the figure. These green turtles dove 772 times on average during the internesting period, and the average dive duration was 16.7–22.6 min.

It is known that during the internesting period green turtles take a rest at the bottom of the sea. From the results of the average of stay depth of each individual, the dive depth tended to become deep while shifting from the dry to rainy season (Fig. 2.10). This indicated that the resting place (depth) of the green turtles is different between the rainy and dry seasons. In the sea area around the Similan Islands, the oceanographic conditions often deteriorate because of monsoons in the rainy season. The deterioration of this oceanographic condition may have changed the depth where the green turtle stayed.

Fig. 2.8 Green turtle installed depth data logger (Courtesy of Dr. T. Yasuda)

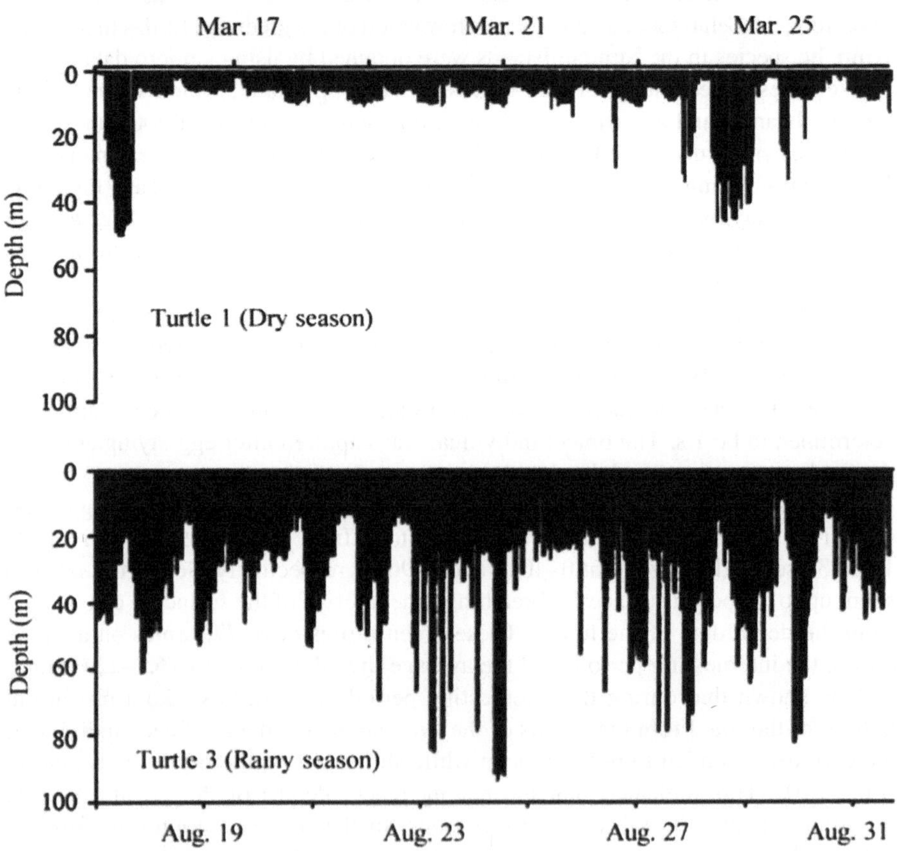

Fig. 2.9 Time-series diagram of depth of a green turtle from the depth data logger during the internesting period (Reprinted from Yasuda (2006) with permission)

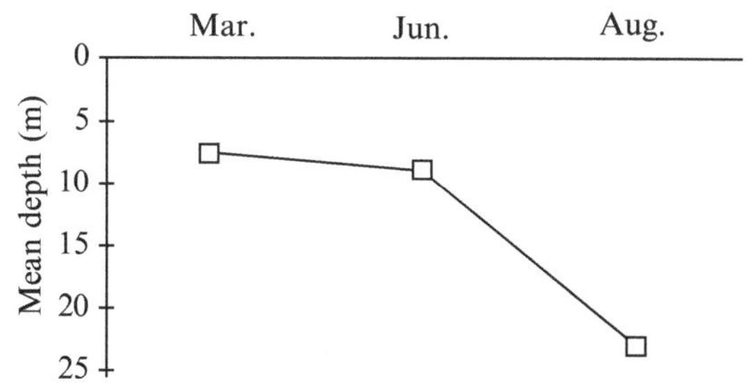

Fig. 2.10 Average dive depth and average dive efficiency of each laying turtle in March and June and August (Modified from Yasuda (2006) with permission)

2.4.2 Homing Behavior Investigation of Rockfish Using Ultrasonic Biotelemetry

Kansai International Airport in Japan opened in 1994 as a real seadrome that took into consideration the maintenance of the environment. The airport island is surrounded by mildly sloped shore protection and on most of these slopes are seaweed beds. The area surrounding Kansai Airport is assumed to function as a nursery for marine creatures because Osaka Prefecture prohibits fishing there. It is known that many expensive fish species such as rockfish, scorpion fish and sea bass live around this area. However, how such fish were using the area surrounding the airport island was uncertain.

It is known that most of the *Sebastes* genus show site-fidelity to their habitat and they return back to their habitat within a few days, even if they were displaced several kilometers away from the habitat. As for the rockfish that live in the shore reef region and the seaweed beds in the Japanese coastal waters, the range of movement is small, and it is known to persist in the habitat. However, whether they return back to the habitat after they are released after transportation as well as other *Sebastes* fish has not been understood. It is important to investigate in detail and to understand such homing behavior when the rockfish's resource is protected and cultured. Then, it was examined using the biologging whether the rockfishes return back to the habitat after transportation and release.

Because the rockfishes inhabit the sea, ultrasonic biotelemetry methods were used. We investigated the sea area around the Kansai Airport Island in the Osaka Bay (Fig. 2.11). An examination operation permit was obtained from Osaka Prefecture and 25 rockfish were captured in three places (A, B, C) of the Kansai Airport island east shore protection area. After 4–5 days, the ultrasonic coded transmitters (V8SC-6 L, Vemco Co., Canada) were installed in the body cavity through a surgical operation (Fig. 2.12). The size of the rockfish's ventral cavity was measured beforehand, and an appropriate transmitter selected. We confirmed the rockfish was not influenced by the transmitter installation. Then, we released them at two points (Fig. 2.11: R1 and R2) that were a maximum of about 4.5 km away from the place of capture. After the release, we performed a tracking study using an onboard receiver (VR28, Vemco Co., Canada; Fig. 2.13).

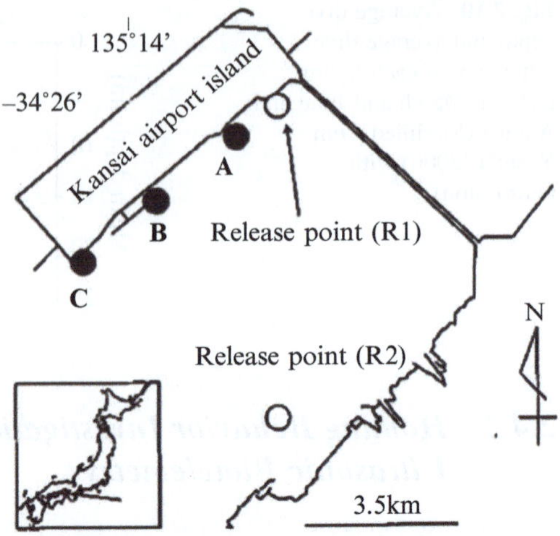

Fig. 2.11 Sea area where rockfish tracking experiments were performed. The ultrasonic transmitters were inserted in the rockfish collected in A, B, and C points. We released them at points R1 and R2 and tracked them (Reprinted from Mitamura (2006) with permission)

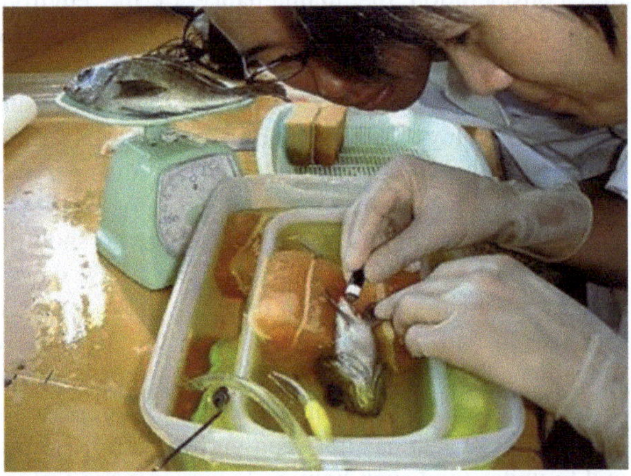

Fig. 2.12 Photograph of the insertion of the ultrasonic coded transmitter in a rockfish under anesthesia (insertion into the body cavity)

Afterwards, 20 rockfish released at the airport shore protection area (R1) began moving along the airport island east shore protection. While these individuals moved from the release along the shore protection area at random within 4 h, when 4 h passed they began moving in the direction of each capture place together. Fourteen individuals returned to the place of capture in several days. From the results, the individual detected the direction of each habitat in 4 h from the release and afterwards they returned back to their habitat. Five individuals that had been released in R2 on the opposite bank migrated along the tidal current just after the release. Afterwards, three individuals returned back to each place of capture within 11 days. The homing rates for point A, B and C were 100% (3 of 3 individuals), 60% (6 of 10) and 67% (8 of 12), respectively. From these results, it became clear that the rockfish returned back to their habitat, as did other *Sebastes* fish when the transportation and release was performed.

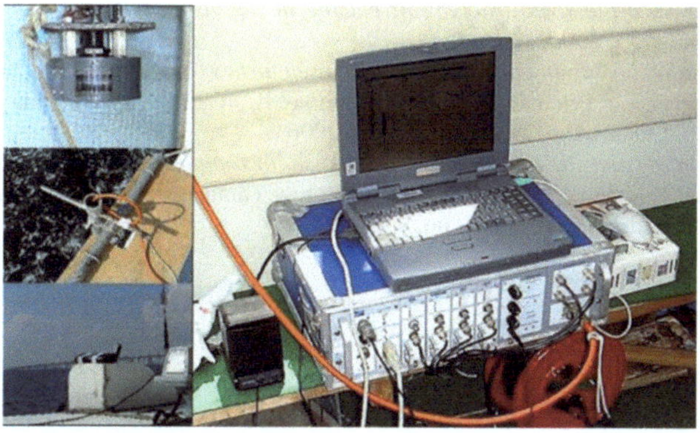

Fig. 2.13 Photograph of tracking experiment using a tracking type receiver (VR28, Vemco Co., Canada). A set of four hydrophones (a) is placed in the water from the gunwale (b). The hydrophone detected the ultrasonic coded signals from the sample fish and the signals were processed with the receiver on board. The direction and strength of the signals were displayed on the PC (d). The position is measured with GPS at the same time (c)

The measurement of difficult to observe animal behavioral information became possible with the biologging techniques, as described above. As for micro data loggers and various transmitters, further miniaturization and increased performance are expected through advanced electronics and communications technology. Moreover, it is expected that micro data logger and transmitters will become less expensive when biologging comes to be widely used and more easily employed as a research technique for various wild animals. We would be very pleased if biologging can contribute to solving social problems through this chapter. Biologging has the possibility to bring further findings by using it together with other surveying techniques, as introduced above. Particularly, combining biologging with remote sensing or geographic information systems (GIS) enables us to analyze animal behavior information, environmental information, and the range of the human activity from an integrated perspective. Moreover, the findings obtained by these methods can become the basic data of further system dynamics analysis and multiagent simulation.

References

Naito, Y. *New steps in biologging science*. In Memories of National Institute of Polar Research, Special Issue No. 58. Biologging Science (ed. by Naito Y.), National Institute of Polar Research, Tokyo, Japan, pp. 50–57, 2004

Rutz, C., Hays, G.C. *New frontiers in biologging science*. Biology Letters 5(3) 289–292, 2009

Cooke, S.J., Hinch, S.G., Wikelski, M., Andrews, R.D., Kuchel, L.J., Wolcott, T.G., Butler, P.J. *Biotelemetry: A Mechanistic Approach to Ecology*. Trends in Ecology and Evolution 19(6) 334–343, 2004

Yasuda, T., Arai, N. *Fine-scale tracking of marine turtles using GPS-Argos PTTs*. Zoological Science 22 547–553, 2005
Yasuda, T. *Migration and diving behaviors of green sea turtles (in Japanese)*. In Aquatic Biotelemetry and Fishing Gear Telemetry (ed. by Yamamoto, K., Yamane, T., Mitsunaga, Y.). Koseisha Koseikaku Co. Ltd., Tokyo, Japan, pp. 76–85, 2006
Mitamura, H. *Homing and site fidelity of black rockfish (in Japanese)*. In Aquatic Biotelemetry and Fishing Gear Telemetry (ed. by Yamamoto, K., Yamane, T., Mitsunaga, Y.). Koseisha Koseikaku Co. Ltd., Tokyo, Japan, pp. 9–21, 2006

Chapter 3
Human Sensing

Yuichi Nakamura

Abstract Places such as homes, offices, workplaces, classrooms, conference rooms, and streets can be considered fields where humans act. The purpose of human sensing is to observe and analyze humans and the social interactions that occur among them in these fields in order to discover human activities and social systems, design and establish new social systems and environments, and develop new information media or artifacts. Technology for sensing humans has been developed in various fields such as medicine/physiology, engineering, psychology, and sociology. Some examples include media processing and artificial intelligence used for observing and automatically recognizing human intentions, human engineering for designing artifacts, and user interfaces (Knapp and Hall 1972; Wickens et al. 2004). By incorporating technologies used in these fields, we will consider how to observe and analyze complicated and multitiered phenomena in target fields as pluralistically as possible. This chapter is organized as follows: Sect. 3.1 describes the type of information to be collected, and Sect. 3.2 explains the details of each sensing technology. Section 3.3 contains examples of acquiring multifaceted data and browsing human activities. These examples demonstrate the latest information media technology. In Sect. 3.4, scenario examples are discussed. Note that sensing technologies for nature and biologging, described in Chaps. 1 and 2, overlap with human sensing technologies. Hence, referring to those chapters is recommended.

3.1 Information to Be Measured in Human Sensing

An example of the experience of learning about farming demonstrates how human sensing would be conducted. Recently, city-dwelling children are being given the opportunity to experience farming life. This form of learning has led some of these

Y. Nakamura (✉)
Academic Center for Computing and Media Studies, Kyoto University, Yoshida-Honmachi, Sakyo, Kyoto, Japan
e-mail: yuichi@media.kyoto-u.ac.jp

T. Ishida (ed.), *Field Informatics*, DOI 10.1007/978-3-642-29006-0_3,
© Springer-Verlag Berlin Heidelberg 2012

Table 3.1 Example of information on humans

(a) External conditions
Bodily movement (see Table. 3.2)
Bodily characteristic (appearance, sweating, body odor, others)
(b) Internal conditions
Physiological condition (see Table. 3.3)
Psychological condition (tension, fear, emotion, comfort/discomfort, others)
(c) Communication conditions
Verbal communication
Nonverbal communication
Interpersonal contact, interpersonal distance, mutual interaction with group

students to eventually undertake practical work, such as cultivation and harvesting. In some cases, learning about farming has even been adopted in the curricula of elementary and junior high schools. Indeed, such an education program provides an opportunity for the students to experience work and promote intergenerational exchanges with local people and to engage not only in agricultural tasks but also with nature. It also provides an important field that encompasses human activities such as defining issues, making and implementing judgments, and obtaining such results as plant growth and harvest. Older students who wish to engage in farming to contribute to society after dropping out of corporate life or taking mandatory retirement may also join experience learning.

For successful learning, it is important to know the ability of the attendees to learn, to analyze the effects of their learning, and to improve learning materials. We also need to examine the students' ability to safely avoid accidents. Another important topic is how the instructors maintain continuous communication with the attendees to hold their interest during the period from planting to harvesting, which sometimes takes several months. Those are important issues on human sensing of observing humans, recording data, browsing and analyzing human activities.

To begin with, we need to determine what type of information concerning humans exists in a given field. Table 3.1 shows a rough classification of the information. First, there are external conditions (bodily movements and their characteristics). By sensing who is moving, where they are moving, and the type of action and bodily movements that are taking place, we can gather information that can be used in planning, evaluating, personal coaching, preventing accidents, and conducting similar activities.

Next, there are internal conditions (physiological and psychological). This study considers what bodily effects or fatigue occurred (physiological body conditions), what interested the subjects and how they felt about it (psychological conditions), and other conditions. Measurement of psychological conditions such as levels of tension, fear, or a sense of burden has often been conducted. Recently, there have been numerous attempts to measure levels of discomfort. Measurement of such physiological and psychological conditions indicates how these fields affect humans in mind and in body.

In addition, communication conditions are also important because numerous humans intricately intermingle within a field. In the case of learning experience about farming, there is a strong correlation between effective communication between instructors and attendees and the ability of the attendees to learn and operate machinery safely. Communication with fellow students helps motivate and incentivize attendees, while the methods used for discussion and decision-making are critical to keep the community in a good condition.

3.2 Sensing Technology

An important component of human sensing studies is to ensure that the system design is suitable for the field to be observed and for the human subjects. In other words, a system that is multifunctional and highly accurate is not always better; various other aspects such as cost, size, weight, invasiveness, and sense of restriction must be considered. For example, contactless and noninvasive sensors are preferable in many cases. Although such sensors are problematic in terms of accuracy, they are often required to establish an environment where the persons being observed can act naturally without a sense of restriction.

3.2.1 Measuring External Conditions

Video recording is suitable for recording visible conditions. Recently, international standards of video compression and transmission (such as MPEG and H.264) have been obtained, and the handling of videos has been facilitated. However, to record fields accurately, devices such as multiple cameras and microphones are necessary. Technological advancements now allow researchers to directly record images and voices using personal computers. Various devices have been developed, for example, to control each video recording to start simultaneously and record the exact time for which each event occurred (time stamp).

However, in general, it is not easy to obtain complete information from videos alone. Table 3.2 shows examples of additional measuring methods that detect complementary information, each of which is explained in the following subsections.

3.2.1.1 Measuring Human Location

Image sensing is often selected for measuring human location and movements because it does not impose any sense of restriction on the humans being observed. When identifying humans, temporal changes are detected and identified by processing image sequences, and targets such as face or body parts are tracked, for example by the particle filter method (Doucet et al. 2001). Automatic tracking has been widely used for pedestrian observation, abnormal behavior detection, security surveillance, and other applications. Various algorithms have been

Table 3.2 Measuring of external conditions

Facial expression:	Image sensor
Movement (location):	Magnetic sensor, image sensor, data glove, motion capture
Location:	Beacon/RFID, GPS, gyro sensor
Sound:	Sound recognition, acoustic recognition
Line of sight/Point of sight:	Eye mark recorder, image sensor

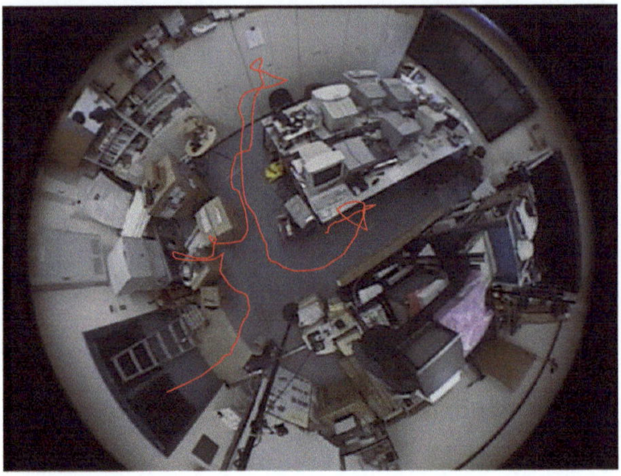

Fig. 3.1 Example of looking down at a room from the ceiling using a fish-eye lens. Movement tracking of a human emerges while observation is shown by the *heavy line*

implemented and widely available, for example, in OpenCV Library (Bradski and Kaehler 2008).

The use of multiple cameras helps researchers observe wide areas. Overlapping the observation areas of those cameras reduces occlusions (areas in which the observed object is hidden by some other object) and enables three-dimensional (3D) measurements. A different method is used to obtain omnidirectional images, as shown in Fig. 3.1, by attaching a special mirror, as demonstrated in Fig. 3.2, or a fish-eye lens.

Other than these image sensors, GPS can also be used for outdoor observations as well as the biologging described in Chap. 2. However, because a portable GPS might cause a measurement error of 1 m or more, there is frequently a need for the previously mentioned image sensors or motion captures, described in Sect. 3.2.1.3, to capture highly accurate measurements of human actions for analysis—that is, measurements that not only determine the location of the person but also with whom or with what that person is in contact.

In addition, if indoors, the location can be identified using devices that facilitate communication over short distances (from 10 cm to about a couple of meters), such as beacons and radio frequency identification devices (RFID). In both indoor and outdoor locations, a device that combines the use of GPS and gyro sensors is often developed.

Fig. 3.2 Mirror to be used to record all-round images. Light from the environment reflects in the direction of descent (camera direction) at the *upper silver mirror*

3.2.1.2 Face Detection, Individual Discrimination

Sensing technologies such as those used in detecting human face, identifying the person, and recognizing facial expression have been actively researched and developed (Ekman 1993). Recently, developments in face detection technology have shown that it is possible to use a relatively low-resolution gray pattern to inexpensively implement face recognition technology into various products including digital cameras and printers. Face detection algorithms are also available in OpenCV Library.

3.2.1.3 Measuring Human Movements and Actions

Human sensing studies often require the ability to recognize and record the types and content of movements in which humans engage. To some extent, it is possible to estimate human intention from their movements. To measure the movement of each part of the human body, various 3D measurement methods (known as motion capture methods) are used. Obtaining 3D measurements with multiple cameras is possible by attaching light sources such as LED markers, which reflect light in all directions, to the human body. Magnetic sensors are also frequently used in situations where it is possible to attach intrusive devices to a subject's body.

3.2.1.4 Measuring Line of Sight and Point of Sight

The locations on which a human focuses his or her gaze are significant for determining the perception condition or interest of the person. This information gives good guidance for designing artifacts and information media. An eye mark

Table 3.3 Example physiological indicators

Electrocardiogram, heart rate, blood pressure, pulse pressure, volume pulse wave, O_2/CO_2 concentration in the blood
Breathing rate, O_2/CO_2 concentration in breath
Electrooculogram, blink, pupil size, focus
Electromyography, evoked electromyogram
Skin potential activity, flicker value, body temperature, facial skin temperature, perspiration
Electroencephalogram, magnetoencephalography (MEG), fMRI (functional magnetic resonance imaging), NIRS (near-infrared spectroscopic topography)

recorder is used to measure line of sight and point of sight. This equipment, involving the attachment of infrared, luminescent devices to a subject's head or glasses, detects the reflection of light on the eyeballs and calculates its orientation. Because eyes are not perfectly spherical, the direction in which light is reflected enables the calculation of the line of sight. However, sometimes it is difficult to calibrate the equipment when it must be worn. Moreover, accurate measurements are difficult to capture if the human subject is distracted by the restriction and feeling of wearing the appliance. To overcome such shortcomings, alternative methods are sometimes used to measure face and line-of-sight orientation by image sensing.

3.2.2 Measuring Internal Conditions

Physical internal conditions such as fatigue and psychological conditions such as fear and tension are also important information to gather in the given field. If studies can detect when human subjects are physically burdened or fatigued, accidents can be prevented and proper operating procedures can be planned. Psychological information is also useful for the same reason because delineating the intention or perception of each person provides key information for defining a problem properly or designing a suitable environment.

Physiological indicators that have been used in medical and physiological fields provide clues to understanding internal conditions. Representative examples are shown in Table 3.3 and most of this information can be measured in real time. For example, it is well known that changes to electrical potential or conductance of skin (including changes to perspiration rates) provide insights into emotion changes. In fact, polygraphs, which simultaneously measure a person's heartbeat, breathing, and changes in skin activities, are used to measure stress or indicate that the subject is lying.

Recently, there have been numerous attempts to explore psychological conditions by directly measuring the brain, as well as experiments to control external machines by reading thoughts from the brain (Dornhege et al. 2007). It has been also reported that cerebral blood flow changes in a certain portion of the brain can be detected with fMRI (described below) when players need to establish trust relationships in a game that requires cooperation. However, such relationships have not been generally explored yet, and future intensive research is necessary.

In addition, to detect muscle usage and fatigue, the electrical potential caused by muscles can be measured. The following section describes neural and muscle activity measurements in detail.

3.2.2.1 Measuring Brain Activity

The electroencephalogram has a long history of being deployed as a method for measuring brain activity. This method measures electrical alterations in accordance with neuronal activity via electrodes attached to the scalp and converts the amplitude of each signal frequency component, indicating brain activity. In general, the higher the frequency of the signal, the more active is the brain. For example, alpha wave signals at frequencies 8–13 Hz show a relaxed rather than an excited state, while frequencies lower than this show a decline in activity (slumber or sleep). Higher frequencies indicate excited activity (indicating that this portion of the brain is activated or tense). Because this method is attractive, it has been used for various purposes. However, it still has serious implementation problems such as the manual skill required to attach electrodes and the noise generated from the muscles around the head, which hamper accurate detection while measuring brain activity during general behavior.

Magnetoencephalography (MEG) is used to measure electrical activity in the brain by detecting magnetic changes. Although the magnetic field arising from neuronal electrical activity is extremely weak, measurement is possible by using a high-performance element with superconductivity (SQUID). Spatial resolution is on the order of millimeters and temporal resolution is on the order of a couple of milliseconds. These characteristics enable highly accurate measurements, although extensive equipment and sensitivity to noise from the environment makes measurement difficult in ordinary situations such as everyday activities.

Functional magnetic resonance imaging (fMRI) and near-infrared spectroscopic (NIRS) topography systems are used as methods for measuring cerebral blood flow and metabolism. The former places the human body in a strong magnetic field and exploits the phenomenon whereby magnetic resonance occurs with the addition of an electric field. Three-dimensional imaging of inside of the human body is obtained by measuring this resonant magnetic field. In particular, because hemoglobin combined with oxygen (oxidized hemoglobin) shows strong magnetic resonance, it is possible to measure any increase in the flow of arterial blood. Although spatial resolution is good, down to a couple of millimeters, temporal resolution is poor, on the order of several seconds. An extensive piece of equipment is required for generating a strong magnetic field, and therefore, it is difficult to take measurements while the subject is engaged in daily activities.

NIRS is another method used to measure brain metabolism. Near-infrared light is projected from a luminescent device placed around the subject's scalp, and reflectance is measured with a different light-receiving device. Near-infrared

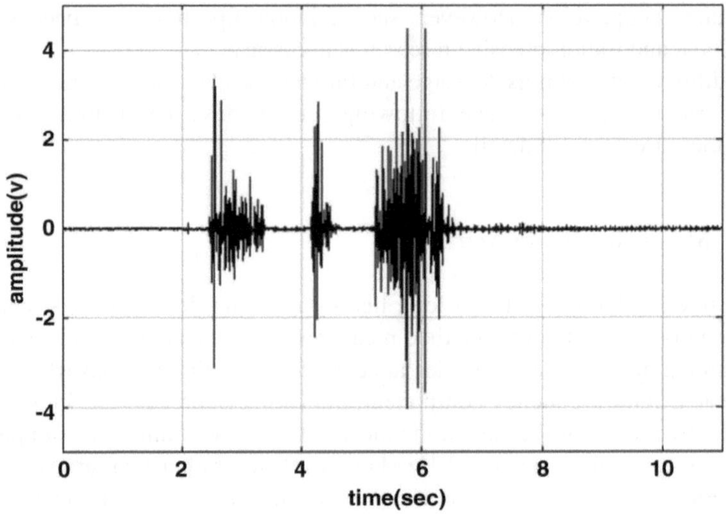

Fig. 3.3 Measuring example of surface muscle potential against the muscle used to move a finger When the amplitude increases, the finger has strength

light at a wavelength of around 800 nm penetrates the scalp and skull and reaches the brain and is absorbed to some extent by hemoglobin. By comparing the difference of absorption between oxidized and deoxidized hemoglobin, the relative concentrations of both can be measured, making it possible to estimate how much oxygen is consumed in each part of the brain. Because measurements can be taken while the subject is active, the use of NIRS has become a relatively popular option. However, it has many shortcomings in the measurement accuracy, for example, a delay in response time of several seconds as well as a spatial resolution of several centimeters.

3.2.2.2 Measuring Muscle Potential

Electromyography (EMG) is a method for measuring potential changes in membranes when muscles contract. The strength of muscle contraction can be estimated by observing the amplitude and changes in electrical potential. When the muscles move, a potential change of zero to several millivolts is detected from the junctions of the motor nerve and membranes, which are mostly located around the thickest part of muscles, and is conducted along the muscle fiber. Surface EMG measures the potential changes on the skin surface, as shown in Fig. 3.3. One advantage of using EMG is that this method can measure muscular contraction even if it does not appear externally. Another advantage is that the signal can be measured before action occurs, because the potential change takes place before the muscles actually contract.

Muscle potential measurement processes have been used in various fields such as medicine, human engineering, and sports science. For example, they are used to develop software interfaces for artificial limbs, for evaluating the usability of artifacts, for measuring the degree of fatigue caused by a work task, and for designing objects or an operational environment on the basis of those measurements. They are also used to help people learn to reduce loads when using their body or to facilitate high performance when engaged in sports. Although they are actively used for various purposes, they still have various problems to overcome, such as the problem of attaching electrodes and sensitivity to noise.

3.2.2.3 Subjective Appraisal and Others

Although various attempts to detect physiological measurements have been successful to some extent, direct measurement of detailed psychological information remains an extremely difficult issue. To gather psychological information, various methods have been employed that ask subjects to verbalize their thoughts or determine psychological states by introspection (protocol analysis (Ericsson 1993)) and subjective evaluation with questionnaires. With these methods, variations occur from one individual to the other, making it difficult to obtain and analyze objective data. Narrative analysis is another method used for recording psychological conditions or experiences from various perspectives (see Chap. 10). In this form of analysis, the focus is on reorganizing an individual's recognition or feelings by reconstructing real experiences as a story or to manifest things not noticed by the individual.

3.2.3 Measuring Communication Condition

The purpose of measuring communication conditions is to determine who expresses what type of information and in what way, or who receives information and in what way, and what type of change is made. The process involves measuring the previously discussed external and internal conditions. For example, if measuring devices indicate that two people are standing facing each other, each of whom is engaged in speech and motion, this affords a clue about who performed what communication. For example, one person (character *A*) could say, "Oh, xx is there!" The action takes place at a certain position and at a certain time with character *B* looking toward character *A*. We need to consider the following as key points.

3.2.3.1 Verbal and Nonverbal Information

The most representative communication is verbal communication, for which hand transcription and automatic speech recognition are often conducted. However, observing and analyzing non-linguistic information is also essential for understanding human communications, because more than 70% of the information that humans send and receive is considered to be nonverbal. Nonverbal behaviors can be various actions such as gestures, facial expressions, the way a person wear dresses, and interpersonal distance.

Conversation analysis is a method that covers not only verbal information but also all other behaviors of the speaker and listener when a conversation is taking place (Psathas 1995). Recently, some projects have attempted to automatically create meeting minutes and support subsequent meetings with the minutes. To record and convey the progression of such conferences, it is necessary to not only record the speech content but also record and analyze nonverbal information including the state of listeners. Recording these types of information requires various forms of sensing and analysis, such as determining which participant fulfilled the main role or ascertaining the atmosphere of discussion, by sensing and analyzing the behavior of speakers, including speech tone and the attitude of listeners, on the basis of factors such as face direction and nodding.

3.2.3.2 Remote Communication

Remote communication using the Internet has become popular to facilitate communication that transcends time and space. Various styles of such communications exist, for example, communication with text media such as emails, message boards and blogs, real-time communication using images, such as video conferences and distance lectures, and communication using educational media such as e-Learning. This chapter considers remote communication that occurs in video conferences as an actual example, because e-Learning is introduced in another chapter.

During video conferencing, images and sounds from each side are transmitted to the other side, promoting real-time dialog. Compared to the example of using only text, video conferencing dramatically increases the amount of information and facilitates various discussions. However, there is often the sense that communication is inhibited compared to dialog in a face-to-face situation. Therefore, to improve communication, we need to measure the extent to which attendees understand the situation of participants in remote locations or how much the participants are paying attention. Remote communication systems should assist the interaction among participants so that important points, which must be focused upon, can be definitively conveyed to remote locations. For example, when an action occurs that indicates or points out something, the indicated object should be conveyed to the other side emphatically.

3.3 Synthesizing Sensing, Data Overview

In previous sections, we described the technologies used for human sensing in a given field, the proper use of which requires the consideration of many factors. As has been already described, these include the proper choice of sensor and sensing

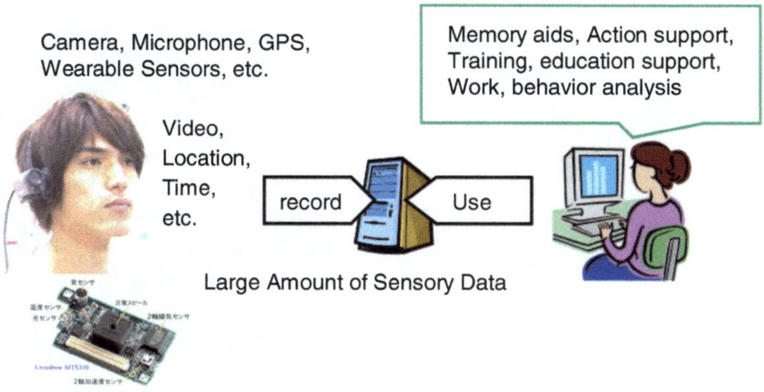

Fig. 3.4 Acquisition and use of the lifelog

method, multifaceted observation, and a synchronous record of obtained data. In addition, to enable flexible browsing, searching, and reuse of data, the use of indexing and an interface (such as a browser) is essential, since the data obtained from various sensors can easily become enormous in volume.

Although meeting such goals in all respects is difficult, we will introduce a method for collecting data using lifelog techniques, providing an index with MPEG7, and browsing with ANVIL as actual examples.

3.3.1 Lifelog (Recording of Individual Acts)

With computers becoming increasingly compact and the widespread adoption of wireless communication, it is now easy to carry a personal computer or sensors and to perform data communication in various fields. Accordingly, we explore the idea of wearing computers and making appropriate services available anywhere.

With such sensors and computers, personal experiences can be continuously recorded as a lifelog and replayed when needed. Lifelogging techniques are expected to be used in various ways to aid personal memory and share experiences among people. The method involves recording and accumulation of videos from a camera attached to the body, as shown in Fig. 3.4; location information (including the use of GPS and gyro sensors); physiological information (sweating, pulse, brain waves, etc.); and environmental information (air temperature, humidity, air pressure, etc.). With advanced computing technology, long-term recording has become possible; for example, a 1 TB hard disk can store 1 month of video data in DVD quality (MPEG2 format of 5 Mbps) assuming 12 h of recording during each day.

However, a problem arises from the fact that the obtained data are long-term and redundant; merely replaying the data will require significant time and effort of the users. One of the solutions is to facilitate rapid access by automatically indexing the

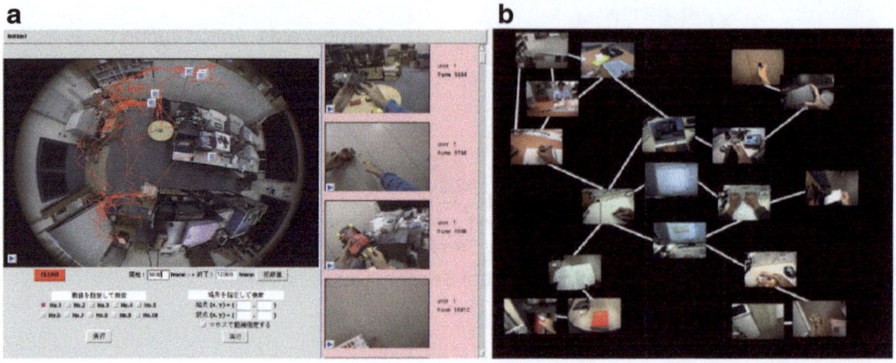

Fig. 3.5 Presentation of lifelog. (**a**) Connection with environmental images, (**b**) Structuring with adjacency

data in terms of location information, action performed, things observed, and similar categories.

For this purpose, the system of Kubota et al. automatically detects scenes in which the user attempted to focus on something, namely the most memorable scene for him or her (Kubota et al. 2002). As seen in Fig. 3.5a, alignment of image data from the user to images taken by a wide-angle camera fixed in the environment provides a comprehensible view for understanding the situation, that is, an overview of what the person was doing and where can easily be observed. In addition, as demonstrated in Fig. 3.5b, obtaining relationships between data fragments based on location, action, and other information and searching and retrieving can be more comprehensible and efficient.

Sumi et al. conducted experiments with multiple people recording lifelog data at various exhibition sites and then analyzed the data (Sumi et al. 2007). They verified that they were able to determine what type of communication the attendees were involved in by using their locations or head directions. They were also able to organize the records and conduct further information provision. These results demonstrate the usefulness of a lifelog in communication sensing.

3.3.2 MPEG7

MPEG7 is a widely used international standard for indexing video content. Because MPEG7 uses the XML schema, it is highly compatible with software that handles XML for various purposes. MPEG7 covers a wide range of content elements such as characteristics of images and audio data, higher semantic descriptions, and metadata such as copyright notices, and allows additional indexing by users.

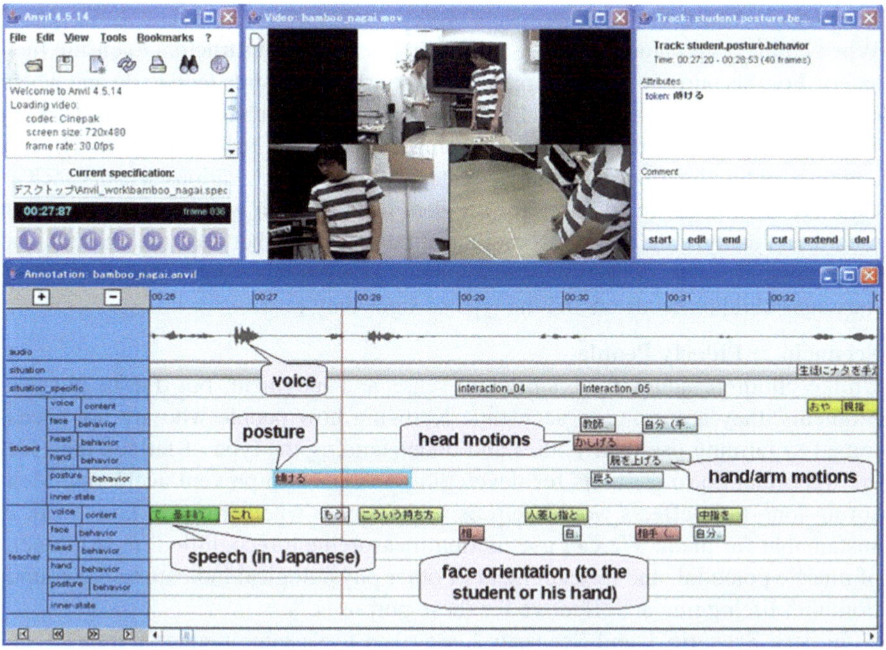

Fig. 3.6 Screen shot of ANVIL

3.3.3 ANVIL

ANVIL (a generic annotation tool for multimodal dialog) is used to give annotations (indices) to multimedia data; it was originally developed as part of a research project about gestures (Kipp 2001). It also utilizes the XML format. Figure 3.6 shows a screen shot from ANVIL. It can also be used as an editor for inserting descriptions at appropriate locations as well as a comprehensible browser. In this example, communication between the teacher (acted by a student) and a student (acted by another student) is recorded with images from three directions, which are then subjected to tagging. The vertical red line at the bottom shows the timing of replaying the videos as they are shown in the upper part. At that time, a posture of "tilt" by the student and the speaking of "well" by the teacher were given. Such indices can be set up arbitrarily by users. Although ANVIL was originally a tool used for tagging, it will also serve as a suitable tool for developing automatic media processing and new applications.

3.4 Scenario Example of Human Sensing

Consider how we can apply human sensing technologies to the scenario of the learning experience about farming.

Scenario 1: Children
When children go outdoors, it is of utmost importance for instructors to ensure their safety. For this purpose, instructors want to know what children tend to do and where or the type of accidents they tend to have. Lifelogging techniques, as described in Sect. 3.3.1, can be a great help for recording the place and time of their movements and how and with what they come in contact. We can attach various sensors such as cameras to children. In addition, for accident prevention, it would be also useful to record and analyze the type of communication conducted between the children and instructors. This data collection is an important element when analyzing the types of messages children focus on and the type of signs they give.

Scenario 2: Elderly People
Unlike children, adults have sufficient sense and judgment, but due to physical problems, they are prone to accidents. Knowing how people who are not accustomed to farming use their bodies and understanding the type of fatigue they feel are important considerations for developing processes to prevent accidents. Measuring long-term effects would help us to determine whether or not farm work is sustainable for them. We can use motion measurement/recognition, measurement of muscle potential, and measurement of other physiological indicators in conjunction with lifelogging are effective for these purposes.

In both Scenario 1 and Scenario 2, we need to sustain attendee's interest and motivation in the long term. While remote communication can be employed, communication patterns must be analyzed to design enjoyable and sustainable experienced learning.

Scenario 3: Designing a Field
Consider a situation of placing instruments or learning materials where attendees can touch them. Attendees have their own experiences and make their new discoveries by looking at, touching, and operating these materials freely. In the case of farming, attendees would have an opportunity to come in contact with and take care of animals. Another case would be learning of instruments and devices in a science museum. To ascertain whether such materials and experiences would be set up well and would work well, it is important to measure information such as who is close to the objects, what time did they make contact, what type of facial expression was shown, and what type of action was taken. For example, we can estimate attendee's interests and attitudes by measuring the location, movement, and the line of sight. If the situation permits, we can analyze how people feel the way of learning by measuring brain activities.

References

Knapp, M., Hall, J. *Nonverbal Communication in Human Interaction*, Wadsworth, 1972

Wickens, C. et al. *Introduction to Human Factors Engineering*. Pearson, 2004

Doucet, A., De Freitas, N., Gordon, N.J.: *Sequential Monte Carlo Methods in Practice*, Springer, 2001

Bradski, G., Kaehler, A.: *Learning OpenCV: Computer Vision with the OpenCV Library*, O'Reilly, 2008

Ekman, P. *Facial expression and emotion.* American Psychologist 48, 384–392, 1993
Dornhege, G. et al. (ed.). *Toward Brain-Computer Interfacing*, MIT Press, 2007
Ericsson, K., Simon, H. *Protocol Analysis.* MIT Press, Cambridge, Massachusetts, 1984
Psathas, G. *Conversation Analysis*, Sage Publications, 1995
Kubota, S., Nakamura, Y., Ohta, Y. *Detecting Scenes of Attention from Personal View Records — Motion Estimation Improvements and Cooperative Use of a Surveillance Camera*, Proc. IAPR Workshop on Machine Vision and Applications, 209–213, 2002
Sumi, Y. et al. *Collaborative capturing, interpreting and sharing of experiences.* Personal and Ubiquitous Computing 11(4) 213–328, 2007
Kipp, M.: *Anvil — A Generic Annotation.* Tool for Multimodal Dialogue, Proc- 7th European Conf. on Speech, Communication and Technology (Eurospeech), pp. 1367–1370, 2001

Chapter 4
Ethnography

Takaaki Tsuji

Abstract Ethnography is a method for describing and modeling phenomena that have occurred in a targeted field. Data collecting through participant observation includes taking field notes, conducting interviews, and having people write self-observation reports. In the analysis of data, discovering the patterns of phenomena which emerge iteratively is important. Describing and modeling the phenomena involves determining and defining phenomena patterns which are discovered and accumulated on a conceptual level and systematically associating them. Ethnography has its origins in cultural anthropology and has developed within social science. The need for ethnography in informatics has emerged because artifacts developed through rapid technological innovations have been introduced in various fields, and induced those involved in the fields to change their activities. Ethnography is a useful method to describe and model circumstances in which people's actions and behavior in a field unfold in a complex relation to artifacts.

4.1 What Is Ethnography?

4.1.1 Definition of Ethnography

Ethnography is a method for describing and modeling phenomena that have occurred in a targeted field. The field may include almost any topic, e.g., the outdoors, work offices, schools, and conference rooms. Ethnography extends into

T. Tsuji (✉)
Graduate School of Informatics, Kyoto University, Yoshida-Honmachi, Sakyo-ku, Kyoto 606-8501, Japan
e-mail: tsuji.takaaki.4v@kyoto-u.ac.jp

T. Ishida (ed.), *Field Informatics*, DOI 10.1007/978-3-642-29006-0_4,
© Springer-Verlag Berlin Heidelberg 2012

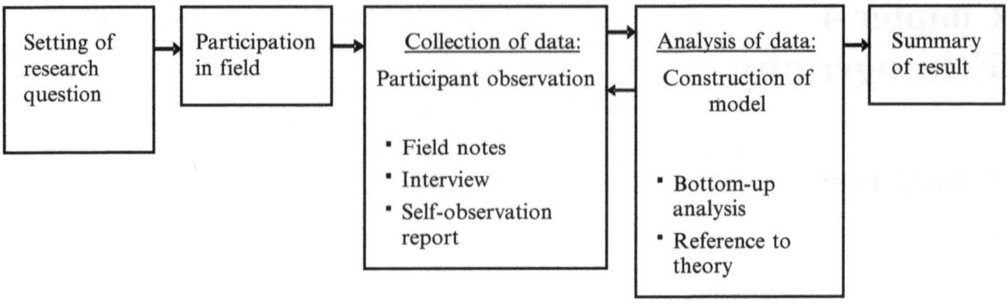

Fig. 4.1 Ethnography process flow

such fields and progresses from there, ensuring the involvement of the people concerned. The process flow of ethnography is outlined in Fig. 4.1. In the process, researchers describe and model phenomena having occurred in the field on the basis of research questions. Here, a "model" is defined as "a single system that inclusively compiles the phenomena and gives them a single settled image." When investigating, the collection and analysis of data are not necessarily independently managed, but progress simultaneously and concurrently.

4.1.2 Ethnography Requirements in Informatics

Ethnography has its origins in cultural anthropology. In the early days of the field, it was used to research "uncivilized" people living in places with unique cultures, such as remote islands in the South Pacific and the African continent. Later, it spread more widely to cover other fields as well. For instance, it was used as a method of clarifying the lifestyle of deviant groups and closed populations in sociology. Today it is actively used in the fields of pedagogy and psychology to illuminate school culture, analyze teacher–student and student–student communication in the classroom, and so on. In this respect, ethnography is a method borne from social science and has developed within that particular field.

The need for ethnography in informatics has emerged because artifacts developed through rapid technological innovations (such as information technology equipment) have been introduced in various fields, and induced those involved in the fields to change their activities. People engage in activities while using artifacts in various fields and being influenced by them. In addition, ethnography is a useful method to describe and model circumstances in which people's actions and behavior in a field unfold in a complex relation to artifacts.

Thus, in the highlighted fields in informatics, the relation between humans and artifacts such as information technology equipment, or that between humans and other humans through such artifacts, is controversial. In such cases, ethnography is used for the following purposes:

- Artifact Design
  Engineers and designers usually design artifacts with certain intentions. In so doing, they use ethnography to verify whether people are likely to use such artifacts as intended. For instance, when a new cellular phone is developed, any gaps in or differences from the original intention are clarified by describing and modeling scenes in which the user is using it. In this way, its design can be improved.
- Field Design
  When a new artifact enters the field, in addition to the human-artifact relation, there are also various circumstances and patterns surrounding the human change. Suppose, for instance, the business manager of a certain company wished to revitalize the sharing of information among employees by introducing the mobile phone as a means of contact. By using ethnography to describe and model the patterns and circumstances of employees using mobile phones to share information, various relations among components such as humans, mobile phones, and others in the office can be effectively designed.

In informatics, ethnography results are used to aid in coming up with suggestions for designing artifacts and the field. In other words, ethnography is used as a design tool.

4.2 Ethnography Process

In this section, I will explain the details of the ethnography process in line with the flow in Fig. 4.1.

4.2.1 Setting of Research Questions / Participation in Field

4.2.1.1 Setting of Research Questions

In ethnography, initially research questions are set. As with many research methods, ethnography involves a need to clarify what to reveal in research, either before research begins or at an early stage of it. Researchers refer to related previous studies and theoretical books, set questions in advance, enter the field, and go into the questions in greater depth. At this time, it is also important to ensure that the research concerns of the researcher and practical requests from those in the field match up.

4.2.1.2 Setting of Informants

Ethnography starts by establishing relations with people in the field. Among such people, those who do related research are called “informants.” As with ethnography, establishing relations with all people is difficult for informants, hence several informants are normally set. Those that are particularly important are known as "key informants." Researchers must establish relationships of trust (i.e., rapport) with informants at an early stage and maintain them from then on. However, this alone is not enough to research informants. The researcher should take account of informants (or key informants) and their surrounding environment. With this in mind, it is important to see the key informants and their surrounding environment.

4.2.1.3 Position of Researchers

Researchers who participate in the field are known as "participation observers." For informants, participation observers are neither complete outsiders nor complete insiders. Thus they adopt a third party perspective, i.e., one that is separate from both. Therefore, researchers should determine their own stance, i.e., the ratio of their participation to their observation in the field. When researchers are focusing on observation, namely when taking an outsider position, they must be devoted to seeing and hearing the phenomena occurring inside, since they are off-site. Conversely, when focusing on participation, namely when taking an insider position, they must enter the site and relate to the practice being carried out there. Investigators should clarify their own perspective according to the purpose of the investigation or requests made by people, and then participate in the field accordingly.

4.2.2 Data Collection Method

The term "participant observation" refers to entering a field to do research on it. It involves learning the meaning and potential structure of phenomena that have occurred in the field by observing or hearing informants. In doing so, researchers make full and effective use of their senses to attempt to discover the patterns of repetitively occurring phenomena in the field. The patterns of both recurring and emerging phenomena reveal an essential part of the entire phenomena. Emerging phenomena occurring in the field can be inclusively described by discovering, accumulating, and systematically relating certain small patterns.

Specified data collecting activities involving participant observation include taking field notes, conducting interviews, and having people write self-observation reports.

4.2.2.1 Making Field Notes

Participant observation starts with making notes concerning personal self-awareness. Researchers observe the practice of informants and their surrounding environment in detail and make notes based on what they have seen. Since these notes may be illegible scribbles and only readable by the person who wrote them, it is important for them to summarize their awareness in their own mind while remembering the 5W1H rule (who, what, when, where, why, and how). Patterns of phenomena that have occurred emerge when memos of awareness are accumulated. The patterns of the discovered phenomena are expressed by sentences and figures prepared following the end of the investigation and accumulated in electronic form as document files. The memos and document files described by those engaged in participant observation are called "field notes." Tools which help in compiling such field notes include audio-visual equipment such as digital video cameras and IC recorders. By rereviewing or listening to the recordings of such audio-visual equipment later, information concerning the field can be reviewed and updated.

4.2.2.2 Conducting Interviews

In participant observation, listening is important as well as observing. This means conducting interviews with informants. Structured, semi-structured, and unstructured interviews with informants are conducted by researchers on a one-on-one basis. In structured interviews the question items are prepared in advance and the interview is conducted outside the field, in a format similar to that of investigatory hearings. In semi-structured interviews the question items are prepared to some extent but informants are also allowed to talk relatively freely. Interviews of this type are also conducted outside the field. Unstructured interviews, however, are conducted without question items being prepared in advance and resemble daily conversation.

The primary purpose of the interviews is to question informants to confirm the validity of discovered phenomena patterns while taking field notes. The second purpose is to obtain clues by which other phenomena patterns may be revealed. Here, it is important for researchers to focus on events that repeatedly recur in informants' answers and words frequently derived from the answers. This makes it possible to obtain clues revealing phenomena patterns that participant observation alone would fail to make obvious. During interviews it is preferable to avoid excessive focus on the responses of a single particular informant or key informant, since visions and perceptions of phenomena often differ among informants. With this in mind, it is important to do some reviewing of responses obtained in previous interviews. Moreover, all interviews must be recorded with an IC recorder or the like, transcribed, and then converted to documents.

4.2.2.3 Having People Write Self-Observation Reports

Field notes as described above involve researchers describing what they have noticed. In contrast, self-observation reports are a means for informants to describe what they have noticed in their own practice. Researchers request informants to write down episodes that reflect their everyday practice and which have profoundly affected them. The episodes should thus include events that occurred which the informants found to be remarkable or extraordinary for their own reasons. In this case, it is important for researchers to closely examine episodes repeatedly described by informants. Self-observation reports enable researchers to acquire details about the informants themselves. They also enable them to obtain clues that confirm the validity of patterns of discovered phenomena based on the aspects of the phenomena, and discover further patterns of the same phenomena.

- Theoretical Sampling

In ethnography, theoretical sampling is used as a method to select the investigation object. This is a procedure involving the selection of an investigation object (e.g., a field or an informant) that improves the validity of the generated model and theory and also continues the data collection. Let us give an example in the specific field dimension. Assuming that in a certain ethnography study the type of student culture of university R is submitted, theoretical sampling means choosing a different university S whose type of student culture has different characteristics and verifying the validity of the type. From the informant perspective, suppose a certain informant T facilitates communication and a pattern to change existing action is consequently introduced. In this case, theoretical sampling means choosing another informant U who is incapable of effectively communicating with members and verifying the validity of U's pattern. In theoretical sampling, patterns of certain occurring phenomena are always compared and systematically related while taking the concepts of commonness and heterogeneity into account.

4.2.3 Data Analysis Method

The data analysis method presented in this section shows the importance of discovering the patterns of phenomena which emerge iteratively when collecting data. Describing and modeling the phenomena involves determining and defining phenomena patterns which are discovered and accumulated on a conceptual level and systematically associating them. First of all, the model should clearly and specifically explain the phenomena of a local field that was actually investigated. At the same time, it should provide abstractness that is also applicable to other fields. Accordingly, the pattern of a phenomenon must be defined on a conceptual level rather than in a raw form. The data analysis method used to describe and model phenomenon is outlined below.

4.2.3.1 Bottom-up Data Analysis

Step 1: The patterns of repeatedly recurring phenomena are discovered and accumulated

The researchers (data analysts) learn the patterns of repeated phenomena in the field and describe them in the form of field notes as sentences and figures. Although the patterns often emerge when the researchers are directly involved in the field, it is also important for them to take note of patterns and review document data after documenting all data collected such as field notes, interview data, and self-observation reports.

- Using a Marker to Discover Patterns

In the abovementioned case, the method of coloring document data with a marker is effective. One approach to this method involves recoating marked items from a light color to a dark one. In this approach, when reading a document for the first time, you mark the parts that you consider as being outstanding by circling them or marking them with a light color. In subsequent readings you phase in the use of darker colors to mark points that you consider to have similarities. This allows you to specify the parts that have commonality and those that differ, and significant markings will stand out as the colors are changed from a lighter one to darker ones. Gradually, for example, you will notice a repeatedly recurring pattern of green → blue → red.

Step 2: The discovered pattern is understood on a conceptual level and the category is generated

As previously mentioned, models generated in ethnography must be abstract to some extent. With this in mind, once the pattern of a phenomenon is determined, its abstraction level is required to exceed the level of the data, and be improved. For this analysis phase, the Grounded Theory Approach serves as a useful tool.

- The Grounded Theory Approach (Glaser and Strauss 1967)

The Grounded Theory Approach (hereafter GTA) is a method involving the development of analysis based on data, extracting concepts from the data, and generating theory via related concepts. Grounded theory is translated into "theory based on data" on the basis of category and properties. "Category" involves the partial or total use of the name to determine the phenomenon to be considered on a conceptual level. "Properties" are conceptual elements which constitute categories and explain the category content indicated by levels such as frequency and intensity.

In step 2 (referring to GTA), after the pattern of a phenomenon discovered and accumulated on a conceptual level is ascertained, the pattern is given a name. This is called "labeling." At the time of labeling, the pattern's name must be based on its characteristics. The pattern of a labeled phenomenon is called its "category."

Step 3: The categories are systematically related and the structure and process of the phenomena are described

After a number of categories have been generated, they are integrated to create a model which includes phenomena. Relating the categories systematically involves describing the structure and process of the phenomena. In this step, it is a good idea to illustrate relations among categories based on circumstances; the KJ method serves as a useful tool for this.

- The KJ Method (Kawakita 1967)

The KJ method was developed by the Japanese anthropologist Jiro Kawakita as a tool to summarize fieldwork data. It targets the generation of new hypotheses by determining the meaningful content from the mountain of data obtained in the fieldwork. In the KJ method, data is first documented. Next, the dynamic structure seen in the documented data is determined. Finally, the data is written out on cards. Classifying these cards and adding captions to them allows the creation of a meaningful set. The relations among the sets are then shown by illustrating them on paper or on a whiteboard.

In step 3, after illustrating the associations among sets and so on, the structure and process of the phenomenon in the text is described in the form of a storyline.

4.2.3.2 Reference to Theory

In ethnography, phenomena are basically described and modeled by analyzing data in a bottom-up manner. However, it is also useful to advance data analysis effectively by using existing theory. Here, the term "theory" includes a scope from general theory, which is sufficiently abstract to be adapted to many phenomena, including local theory, which specifically explains respective phenomena. By referring to a category (concept) emerging from bottom-up data with one emanating from theory in deductive logic, it becomes easy to name concepts and also relate them.

The processes described in Sects. 4.2.3.1 and 4.2.3.2 are summarized in Fig. 4.2.

- Triangulation

The analysis of data in ethnography is considerably dependent on the interpretation of individual analysis. Therefore, when indicating analysis results, it is important to provide a presentation superimposing multiple kinds of information, rather than only one. This is known as triangulation. This technique makes it possible to improve the certainty of the analysis result interpretation.

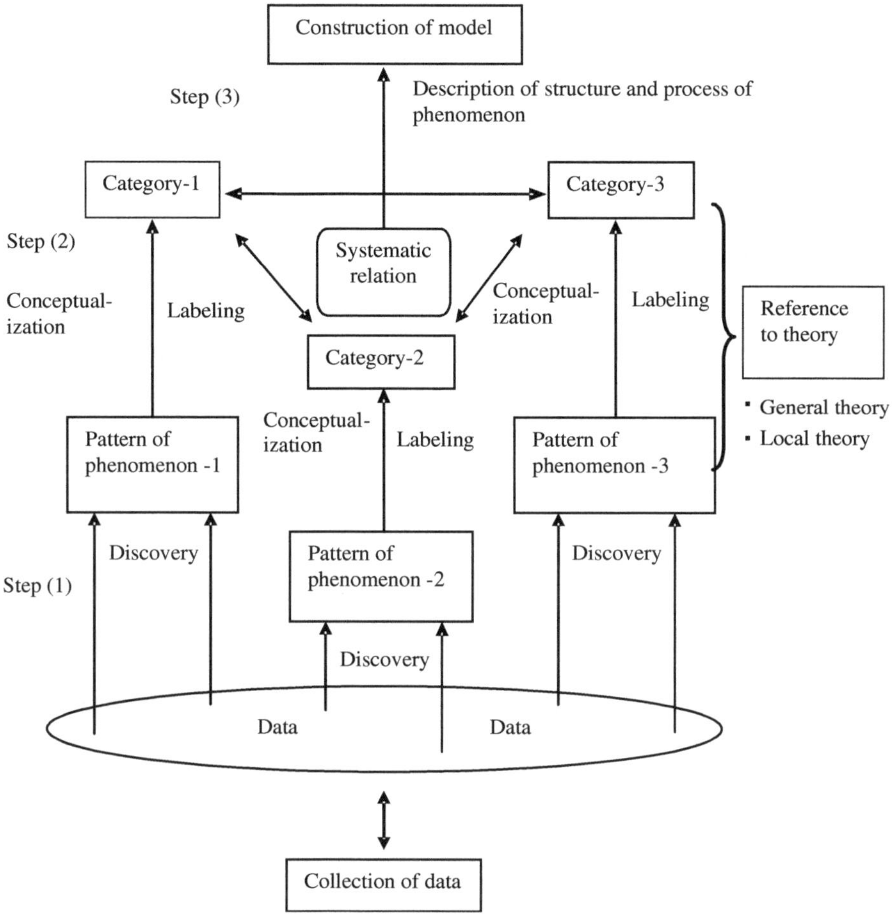

Fig. 4.2 Data analysis process

4.2.4 Summarizing Results

Ethnography can be summarized in the form of case studies. Cases are conveyed to those in the field and to researcher communities. In ethnography in particular, the kind of language used to summarize results is important. The results can then be conveyed to a researcher community in the form of articles, using theoretical terms. However, bearing in mind that the results are also conveyed to the field, the summary must be in language people on-site can understand. In the case of informatics, it is particularly desirable for engineers and designers who design artifacts, or those in the area of the designing site itself, to be able to utilize the results. For this reason it is important for researchers and analysts to describe results in language that facilitates communication with such persons.

4.3 A Case of Ethnography Usage

This chapter describes an example of using the ethnography process, in particular describing information gained as part of a collaborative course between Japan and America. Recently, with the development of the Internet, university classes have featured the practice of remote cooperative learning connecting Japanese and overseas universities. Here, connecting universities A and B in Japan and America respectively via the Internet, I establish a learning environment where students of both countries discuss on a bulletin board system, focusing on the English learning of Japanese university students. I use ethnography to describe and model the learning process and consequently obtain suggestions concerning the design of the learning environment. Note that with respect to data examples and expository writing, I quote from Tsuji et al. (Tsuji et al. 2007) as appropriate.

4.3.1 Outline of Practice

Three Japan–US mixed groups comprising a total of five people (two learners on the Japanese side and three American learners) were organized. I requested the learners to hold free discussions on a specific theme concerning bioethics on the Internet. As media for conducting the discussion, I used the courseware of the American Black Board System, which is an Internet-based study control system. The language used was English. The implementation period was about 2 months, during which time class was conducted every Wednesday. For each of the three groups, I set one Japanese student as a key informant: students J, K, and L. These individuals were provided with sufficient explanation concerning the practice, and all of the group members agreed to cooperate in the study (Fig. 4.3).

4.3.2 Data Collection

Concerning the three key informants, I collected data as follows:

I observed the activity of key informants during the class hour, wrote down points I had noted, and wrote his own field notes giving details of their words and actions.

I provided the key informants with self-observation reports and asked them to describe the details of the content and type of work they conducted, as well as their awareness at the time, in as much detail as possible, both during and outside class hours. Each week, I provided him with a self-observation report form for a 1-week period (one A4 sheet per day, seven sheets in total), kept it on file, and collected it on the class day (Wednesday). I repeated this cycle weekly during the implementation period, a total of 9 weeks.

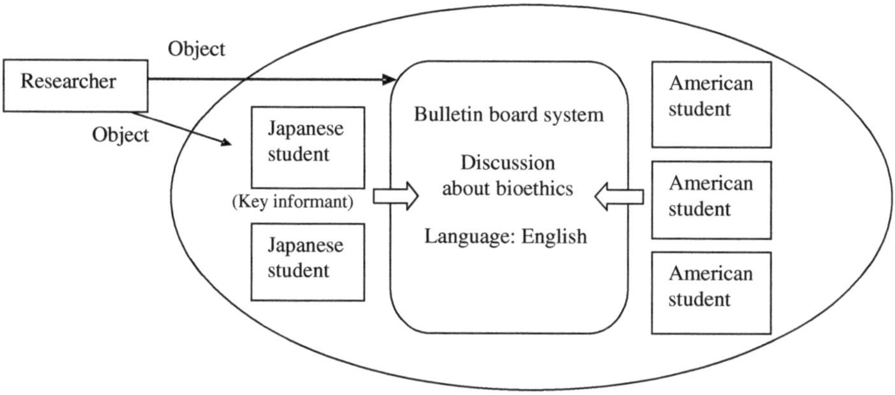

Fig. 4.3 Composition of one group

I collected the key informants' self-observation reports prior to class every Wednesday and examined them. At the end of each class hour, he interviewed the key informants for 30–60 min in a private room. Thus, nine interviews were conducted in total. All interview sessions were recorded with a digital micro recorder, computerized in line with the protocol, and compiled.

I also collected communication logs between Japan and America on the bulletin board system of each group.

4.3.3 *Data Analysis*

Here I explain the results of analyzing the data obtained for the first, middle, and final terms of practice. I then explain the repeatedly recurring patterns that emerged in each process and categories generated based on the same, presenting data examples.[1]

4.3.3.1 First Term

- Repeatedly Recurring Patterns[1]

  Data Example: I reread a self-introduction post I had sent to an American student the other day. Perhaps because I wrote it in a hurry, some of the sentences were grammatically wrong and it seemed difficult even for me to understand them. I believe my English reading ability is reasonably good, as I had a

[1] Self-observation reports from the 1st to the 72nd day are described. As an example, the self-observation report from the 1st to the 10th day is referred to as self-observation report 10. Nine interviews are described; for example, the first interview is referred to as interview 1.

number of opportunities to read the language at university. However, my exposure to English composition was limited, although I thought at the time I was focusing quite heavily on it. From here on I intend to concentrate on my writing, sticking to grammar and diction (tense, syntax) as much as possible. From the American students' responses, I hope to find out how effective studying English for the purpose of passing exams is. (From the self-observation report 2 of student K.)

Data Example: I teach English to first-year high school students once a week for about 3 h, but the lessons are good for me as well since they force me to recall things I had forgotten. Today we studied participial construction, and as I thought about it I realized that participial constructions are actually used for real communication purposes and I reviewed the subject with enthusiasm. High-school English in Japan is stigmatized in many ways (e.g., it is often derided as 'examination English'), but I believe it teaches many key elements as a basis for learning the language. (From the self-observation report 10 of student J.)

Data Example: Today, I checked my email at school, printed out the emails I had received, brought them home and read them. After all, it's still difficult for me to look at English shown on a PC screen and translate it into Japanese in my mind at that moment so that I can understand it. I just feel that I can better understand it if I have time to read it carefully on paper and make notes on it. (From the self-observation report 20 of student J.)

Data Example: From the time the American students post something until the time we respond to it, we always carry out a specific process sequence. First of all, we print out the posted texts on a bulletin board system and read the hard copies. This is because Japanese students are used to reading, underlining and sectioning English texts so we can remember them more easily before summarizing them in Japanese in our minds and comparing them with what we had previously imagined they meant. We then summarize our opinions and post them on the American side. In other words, when we think about what English sentences mean, we switch from English to Japanese mode. We also learn not from what is on the screen but from what is on paper. In short, it can be said that even when the learning environment is unfamiliar, we learn by spontaneously switching to the learning style we are used to. (From the self-observation report 18 of student K.)

Data Example: My work flow up to the time of posting: (1) Consider the points I want to make and post. (2) Write out the points, itemizing them in Japanese. In this step, instead of making a Japanese translation I draw a relationship diagram to facilitate organization. (3) Compose the English sentences on paper, using a dictionary and reference books. (4) Recheck the sentences and post them on the bulletin board system. In step 2, it's relatively easy for me to consider the points I wish to make in my composition by summarizing them and not translating them completely into Japanese. Today, in step 3, what I checked in the reference book was the usage of "some others..." and type 2 conditional sentences. (From the self-observation report 31 of student K.)

In the first term of this practice, in reading and writing exercises done on the bulletin board system, the Japanese students used a learning strategy based on their past experiences in using English. Namely, they prioritized the use of errorless grammar and wording and read and wrote in a way similar to that which they did in taking their university entrance examinations.

Generation of Category 1

I labeled the abovementioned way the students used English as "English use in an examination culture." Subsequently, I labeled the attitude of students prioritizing and valuing the use of flawless grammar and diction and the proper use of syntax as "adherence to English use in an examination culture." Likewise, I labeled the manner of English composition and reading as expressed by students in the form of a calm and familiar learning style as "a learning strategy based on examination culture."

4.3.3.2 Middle Term

- Repeatedly Recurring Patterns 2

  Data Example: In taking university English exams, I had thought that the idea was to read the intent of the person who wrote the question and in answering it to use this or that syntax, this or that expression, and to apply the knowledge I had accumulated appropriately. I now understand that you must also think of the questions themselves. I had thought this would be easy but sometimes it is quite difficult. (From the self-observation report 31 of student K.)

  Data Example: Up to now I have not felt comfortable in answering questions unless I remember to use the proper syntax and so on. However, the postings of people on the American side are rather difficult to understand if you try to understand them in strict grammatical terms. So I feel it is rather different from reading the questions in a university examination. (From the interview 4 of student K.)

  Data Example: For example, in English examinations, specific terminology is always included in notes at the bottom, right? Today I felt that using the given technical bioethics jargon properly would be problematic at best. (From the interview 3 of student K.)

  Data Example: For difficult words, I have to write down the Japanese meaning or I will be unable to understand the meaning of the sentence properly. In the worst cases, I sometimes have to refer to the dictionary a number of times, especially for medical terms. (From the self-observation report 20 of student J.)

In the first half of the middle term of practice, we obtained statements on the basis of which I were able to sense how students who were used to the "examination culture" were feeling confused and very aware of their inexperience in reading and writing practical English texts. Careful examinations of the data obtained repeatedly confirmed that these feelings arose as a result of their interacting with students on the American side via the electronic "Network in English" system used in this study.

- Generation of Category 2

I labeled the abovementioned feelings of confusion and inexperience emerging from the students as the "conflict between an examination culture and a practical culture."

- Repeatedly Recurring Patterns 3

Data Example: The other side seems motivated. Recently we even discussed festivals held over there with them. It's starting to feel like we are actually communicating. It feels, well, good somehow. It feels relaxed and casual. (From the self-observation report 6 of student J.)

Data Example: I think my resistance to or worries about the learning format are certainly easing, because each time I am reminded that even simple tests can be understood by the other side. (From the interview 5 of student J.)

Data Example: Today there was a posting from the American side. It was a response to a message we had sent the other day. It said "We'll be kind." In other words, since the posting we had sent was in poor English, they used simple English for us. To be honest, I was a bit shocked, but it can't be helped as that is just the way it is. (From the self-observation report 11 of student L.)

When the latter half of the middle term commenced, we observed a phenomenon whereby students J and K overcame the abovementioned conflict by communicating smoothly and effectively with the American students. On the other hand, it emerged that communication between student L and the American students did not go well, hence the conflict remained unresolved for that student.

- Generation of Category 3

I labeled the communication status with the abovementioned students on the American side as the "relationship with American students."

4.3.3.3 Final Term

- Repeatedly Recurring Patterns 4

Data Example: This came to mind reading my recent posting, but I think the expressions, syntax and wording I use have been getting fixed to some extent. In terms of word order, I tend to always extract the same things from my overall knowledge of English expressions to some extent. One of my partners in the group was saying that he had recently gotten faster in writing English texts in BB than before. This may be because his English ability had improved, but I think it may also be that for him as for me, the expressions and syntax to use have become fixed to some extent. (From the self-observation report 47 of student K.)

Data Example: One problem is the wording to use when you request something from someone else. For example, in high school I heard many times that in making a request, "Could you" can often be replaced by a simple "Please." But in cases like this the options to use are fixed to some extent, and although this particular example involves making requests, for other cases the options are also mostly fixed. (From the interview 6 of student K.)

Data Example: In today's class, I sent a reply to the American side. It took a while to think of what I wanted to say, and I was again slow in translating into English. As a first for me, I tried using a translation site, and despite some differences between what it said and what I wanted to say, I thought it was useful. However, I wish its translations were more accurate. (From the self-observation report 13 of student L.)

Data Example: When creating the content of (the report of the) web page, I wanted to put the Japanese content that appeared in the book written by teacher V, using software to help me translate it into English. However, translation software is not at all that reliable and it sometimes puzzles me. (From the interview 7 of student L.)

Data Example: With a translation site, I was able to work out English meanings without any of the trouble I had previously had. (From the interview 3 of student K.)

Data Example: Well, English is difficult for me. However, today I tried using a translation site and I think it helped me work out certain meanings. (From the interview 2 of student K.)

During the final practice term, it became evident that the students had changed the learning strategy they had previously used in reading and writing English texts. We frequently observed the phenomenon of students J and K imitating the vocabulary patterns of the American students, i.e., using their sentence expressions in writing or reading. On the other hand, I repeatedly observed the phenomenon of student L using a web translation site in writing English text.

4.3.3.4 Reference to Theory

Situated learning (Lave and Wenger 1991) is one theory to explain the process of students participating in a certain community of practice and using oddly located tools and resources to structure their learning activities. Here, "community of practice" means a place where social practice enables the acquisition of knowledge and skill through group participation. In the theory of situated learning, "resources to be referred to for use each time according to learning scenes" are called "resources." Such resources are regarded as sociocultural resources of the community of practice, through which students acquire sociocultural aspects and beliefs inherent to the community.

Recently, in addition to real space, collaborative learning scenes on electronic networks such as CSCL (Computer-Supported Cooperative Learning) have also emerged as communities of practice. With this in mind, groups of Japanese and American students involved in this practice can also be regarded as a certain kind of community of practice.

In this case, Japanese students referred to resources such as American students' patterns of vocabulary or translation software on a web site. Moreover, the use of such resources was found to affect their value and belief systems.

- Generation of Category 4

I labeled the learning strategy of the abovementioned students J and K as "the use of American students' vocabulary patterns as a resource," and the learning strategy of student L as "the use of translation software on a web site as a resource."

- Repeatedly Recurring Patterns 5

Data Example: A lot of the detailed grammar they make us learn in school is probably far beyond what we need. Actually, we can liaise with CWRU students without using it so much. (From the self-observation report 47 of student K.)

Data Example: In my self-observation report for the first week, I wrote that I wanted to stick to English for the purpose of passing examinations in this project. During the project, however, my main concern became one of striving to build relationships of trust with the other side. I feel that I was able to get used to communicating in English in a more flexible manner than I had thought I would be able to. (From the self-observation report 65 of student K.)

Data Example: It just took time for me to translate things into English. I found writing consistently good sentences to be unexpectedly difficult. Since I wanted to minimize the number of mistakes in my writing, I made considerable use of a translation site. However, I think this is why it took me such a long time. I really feel my lack of English ability. (From the self-observation report 70 of student L.)

While being introduced to this aspect of the theory of situated learning and analyzing data obtained in this collaborative course, I focused on value and belief systems concerning the students' use of English. As a result, it emerged that students J and K, who used the American students' vocabulary patterns as a resource, had obtained a value and belief system for English use resembling practical culture. Conversely, for student L, who used a translation site, the student's mention of "writing consistent sentences, writing without making mistakes" makes it clear that the student retained the use of a value and belief system for English use close to that of the "examination culture."

- Generation of Category 5

I labeled the value and belief system for the English use of students as a "perspective on English" and decided to call changes made to it "reconstruction of a perspective on English." Moreover, concerning the "adherence to English use in an examination culture" in category 1, I labeled it as a "perspective on English in an examination culture."

4.3.3.5 Summary

Systematically relating the abovementioned generated categories 1–5 and summarizing the structure and process of English learning of the students on the Japanese side, I can summarize them as follows:

In this research, I analyzed Japanese students' English learning in an electronic learning environment where they were asked to use English practically using ethnography. Consequently, the following results emerged concerning Japanese university students. (1) When reading and writing, Japanese students used a learning strategy based on their past experience of English use in an "examination culture." (2) Japanese students faced a conflict because of the difference of English use between an examination culture and a practical culture. (3) In the process of adjusting to a practical culture, Japanese students referred to resources such as American students' patterns of vocabulary or translation software on a web site. The students who were capable of effectively communicating with American students used the American students' vocabulary patterns as a resource; those who were incapable used translation software on websites as a resource. (4) By making use of these resources, Japanese students transformed their learning strategies and perspective of English, which had been based on an examination culture. These results suggest that English learning in an electronic learning environment is interwoven into the historical context of students' usage of English and the relation of such usage to situational context such as other persons and resources in the environment and those influenced by characteristic of the technology used in the practice.

4.3.4 Suggestions for Designing Learning Environments

On the basis of the abovementioned knowledge acquired, I here present suggested points concerning the design of learning environments.

- Uneven Distribution of Resources in the Learning Environment

The results obtained in this study indicate that the use of resources has a major impact on learning trajectory. The study also revealed that students facing a conflict situation were able to use available resources to overcome it. From this it can be concluded that, in designing a learning environment within an electronic network, it is important to distribute the resources that students can use and access in advance to encourage their active participation within the environment.

- Intervention by Teachers and Teaching Assistants

Two forms of trajectory emerged in this study regarding the use of English by students. It is important to use the ethnography results obtained in this study to monitor the circumstances of students, determine their trajectories, and support teachers and teaching assistants (TAs) in their efforts to exert a positive influence on them. Of particular note in this study is that certain discrepancies occurred that subsequently led to differences in the way the students of both countries communicated with each other. Therefore, to avoid allowing such discrepancies to occur, teachers and TAs must intervene in the students' practice as necessary and control their learning trajectories to keep them within a certain scope.

References

Glaser, B.G., Strauss, A.L. *The Discovery of Grounded Theory: Strategies for Qualitative Research*. Aldine Publishing, Chicago, Illinois, 1967

Kawakita, J. *Abduction — For Development of Creativity*. Chuo Koron Shin Sha, Tokyo, Japan, 1967

Lave, J., Wenger, E. *Situated Learning: Legitimate Peripheral Participation*. Cambridge University Press, New York, 1991

Tsuji, T., Nishimura, S., Nojima, E. *A Situated Approach to English Learning in Distance Collaborative Course Which Connects Japanese Students and American Students*. Journal of the Japan Society for Educational Technology 30(4) 397–407, 2007

Chapter 5
System Dynamics

Kazuyuki Moriya

Abstract Recent social and economic developments have raised concerns regarding various social issues such as those involving the environment, food, depopulation and aging, and garbage. Environmental and food problems occur on a global scale, while other issues are considered regional. It is very difficult to design investigations to clarify possible causes and solutions because these social issues are derived from various complex human activities. To solve these problems, we must elucidate their mechanisms and clarify the relationships among various events in the field. For these purposes, an approach that investigates the causes of social issues by considering society in the form of a model and connecting causal relationships among events can be effective. We can thereby understand society through the behavior of a social model and can predict future situations. System dynamics is a computer simulation technique that is highly useful for understanding society based on logical thinking. In this chapter, the concepts of system dynamics and its applications are introduced.

5.1 System Dynamics as a Simulation Technique for Social Systems

System dynamics (SD) is a numerical simulation technique that was developed in the late 1950s by J.W. Forrester of the Massachusetts Institute of Technology. Forrester applied the method of system analysis used in engineering to dynamically analyze systems in the fields of business administration and social science. In his book *Industrial Dynamics* (Forrester 1961), Forrester analyzed corporate activity, while in *Urban Dynamics* (Forrester 1966), he applied SD analysis to urban

K. Moriya (✉)
Graduate School of Informatics, Kyoto University, Yoshida-Honmachi, Sakyo-ku, Kyoto 606-8501, Japan
e-mail: moriya@i.kyoto-u.ac.jp

T. Ishida (ed.), *Field Informatics*, DOI 10.1007/978-3-642-29006-0_5,
© Springer-Verlag Berlin Heidelberg 2012

planning. Finally, Forrester developed SD into a mature simulation tool that could represent large-scale social systems.

In *The Limits to Growth*, published by the Club of Rome in 1972, the authors tried to predict the condition of the world after 100 years using Forrester's "World Model," an SD-based model that considered the world population, industrial investment, natural resource use, and environmental pollution. Unfortunately, readers of this work focused more on critiquing the numerical results and did not recognize the usefulness of SD as a numerical simulation methodology. At that time, there was generally little consideration of the limitations and problems of large-scale simulations in social science. However, advances in graphical-interface software for personal computers (e.g., STELLA, Vensim) have contributed to increased use of SD since the 1980s. Meadows et al. published the revised edition, *Beyond the Limits* (Meadows et al. 1992), in which they discussed global development over the first 20 years in the scenario of Limits to Growth. In 2004, they published *Limits to Growth: The 30-Year Update* (Meadows et al. 2004) in which they summarized some of the relevant data and the insights they have acquired over the past three decades. Recently, SD has attracted attention as a tool for investigating system characteristics rather than for generating predictions based on numerical simulations. In addition, the use of SD as a learning tool for system thinking in elementary and secondary schools is spreading in Europe and the United States.[1]

SD offers an effective tool for simulating and analyzing complex systems with various interrelated factors and has a wide range of applications. For example, say we want to analyze an environmental issue. With SD, we build a social model that includes factors related to ecosystem and human activities, focusing on the flows of indicator substances (such as carbon dioxide, a causative factor in global warming). By executing this model under SD, we can grasp the relationships between factors and the flows of indicators through both numerical results and graphical visualization. In addition, SD modeling can aide in decision making and policy examination because the results of various scenarios can be easily compared.

For example, Liu et al. (Liu et al. 2002) developed a nitrogen cycle model for a system integrating beef cattle production with rice cultivation. Using STELLA, they assessed several scenarios for recycling rice byproducts (rice straw) and livestock waste (animal manure) coming from each sector of rice cultivation, beef cattle production, and the extra nitrogen pool. They showed that recirculating production and a type of complex recirculating production in rice paddies and forage fields were superior to ordinary production for reducing total nitrogen emissions, increasing the manure restoration rate, and elevating the feed self-sufficiency rate. Thus, SD can also be applied to the analysis of real production systems.

[1] The Massachusetts Institute of Technology has been developing the Institute of Technology System Dynamics in Education Project, whose Road Maps provide a detailed explanation on SD from its basics. These materials are very useful for beginning learners in SD.

To develop SD models of social systems, as much useful information as possible needs to be collected. Observation and descriptive techniques, such as remote sensing and geographic information systems (GIS), biologging, human sensing, and ethnography, are particularly useful for gathering information on the systems involved in environmental issues. For instance, to develop a conservation system for endangered marine animals, it is necessary to obtain behavioral and ecological information about the target animals. In this case, an effective method of information gathering (e.g., biotelemetry) must be selected. As another example, to develop a local environmental conservation system, we should obtain information about the state of land use in the region using remote sensing and GIS techniques. If we want to include human awareness and activities into the system, knowledge obtained from human-sensing and ethnography will provide good references for developing a model. Thus, to develop an analytical model, observations and descriptions in the field are required. Meanwhile, SD simulations can provide observations of the system's behavior along a time scale. Observing the state of factors and changes in their relationships in the model is useful for identifying problems in the system. SD simulations can also be used to predict future conditions. For example, using Forrester's "World Model," SD has been used to predict changes in the world population, trends in industrial investment, utilization of natural resources, and environmental pollution 100 years in the future.

In this chapter, we use STELLA (isee systems, inc.) to execute some examples and describe models. STELLA is one of the most popular SD software packages and it supports a very intuitive graphical user interface that allows beginners to easily learn about SD. Vensim, which is free for educational use, has functions equivalent to STELLA. Descriptions of these software products and an introduction to SD are available from Road Maps in the System Dynamics in Education Project detailed in footnote 1 as well as in the references listed at the end of this chapter.

5.2 Stock, Flow, Converter

In SD, a system model is described by stocks, flows, and converters. SD is the leading technique for observing and predicting changes in the state of elements and flows of objects and information among elements along a time scale. A stock can be thought of as a container in which something is stored. A flow can be considered a valve that controls the inflow to or outflow from a stock. A converter, which is another element distinct from stocks and flows, is used for defining auxiliary variables and constants (Deaton and Winebrake 1999).

The bathtub model can illustrate the concepts of stocks and flows. This model consists of a bathtub, a faucet, and a drain, and is often presented in introductory texts on SD. The state (water volume) in the bathtub changes constantly according to the relationship between the inflow rate from the faucet and the outflow rate to the drain. As such, the bathtub corresponds to a stock. The term "level" is often used instead of stock. The faucet and drain correspond to flows because the inflow rate

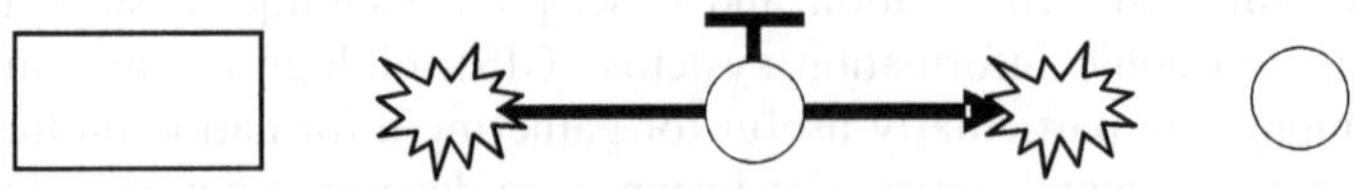

Fig. 5.1 Symbols representing a stock, flow, and converter

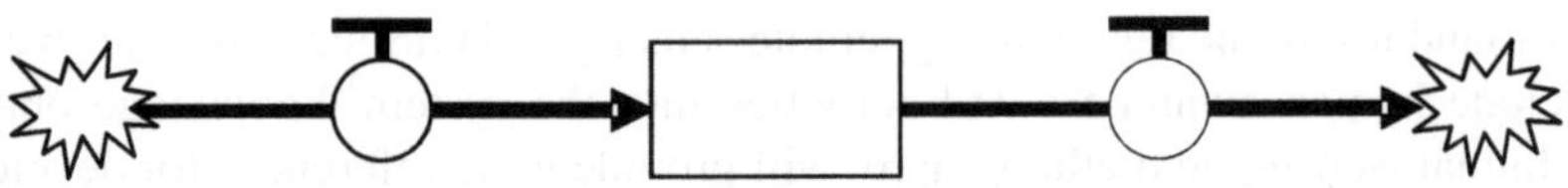

Fig. 5.2 Bathtub model

from the faucet and the outflow rate to the drain change the state (water volume) of the bathtub. The term "rate" is often used instead of flow. In this chapter, we use the terms stock and flow according to the STELLA definitions. Figure 5.1 shows the symbols for stocks, flows, and converters that are used in this chapter.

A diagram of the bathtub model is presented in Fig. 5.2. The symbols on the left side of the flow faucet and right side of the flow drain represent the model boundaries.

To execute the model, concrete values for stocks, flows, and converters, as well as relational equations among them, need to be set. To set these values, data collected from statistics or observations are usually used. However, abstract values and relational equations can be used for conceptual models, as explained in Sect. 5.4.

For example, if we execute the Bathtub model with the flow Faucet = 5 l/min, the flow Drain = 4 l/min, the initial value of the stock Bathtub = 0 l and a time interval (Δt) of 1 min, the stock Bathtub will increase by 1 l every minute. This relationship can be described in SD as (5.1):

$$\boldsymbol{Bathtub}(t) = \boldsymbol{Bathtub}(t - \Delta t) + (\boldsymbol{Faucet} - \boldsymbol{Drain}) \times \Delta t \tag{5.1}$$

where t represents time and Δt a time interval.

In STELLA, if we only connect the stock and flows as shown in Fig. 5.2, an equation such as (5.1) is automatically generated.

5.3 Feedback Loop

In SD, the flow of information among elements can be described by connecting stocks and flows to one another using connectors. These connectors allow for the easy creation of models that include feedback loops. First, we will explain the

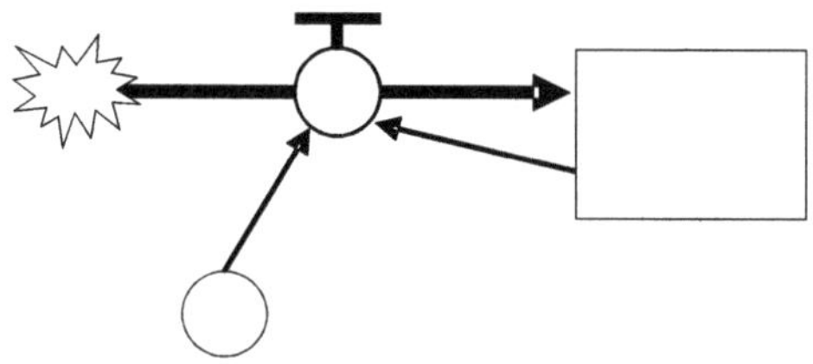

Fig. 5.3 Savings model based on compound interest

structure called a positive feedback loop in a so-called savings model based on compound interest. Figure 5.3 presents the system diagram of this model.

The model shown in Fig. 5.3 consists of three elements: Deposit Account (stock), Interest Income (flow), and Interest Rate (converter). These elements are connected to each other by connectors. A connector is used to indicate the flow of information fcrom one element (source) to another (destination) using an arrow. Unlike in the Bathtub model in which we directly set the flow Faucet to a static numerical constant (5 l/min), here we set the flow Interest Income to the equation "Deposit Account × Interest Rate." In the model shown in Fig. 5.3, we set the value of the converter Interest Rate to 0.05 and the initial value of the stock Deposit Account to 10,000.

The equation for the stock Deposit Account in Fig. 5.3 is then as follows (5.2):

$$\boldsymbol{Deposit\,Account}(t) = \boldsymbol{Deposit\,Account}(t - \Delta t) + \boldsymbol{Interest\,Rate} \times \Delta t \tag{5.2}$$

When we set the time interval (Δt) to 1 year and run the model, the change in the stock Deposit Account up to 2 years later is as follows:

After 1 year,
Deposit Account (1) = ***Deposit Account*** (1-1) + Interest Income × 1
= ***Deposit Account*** (0) + Interest Income
Interest Income = the state of stock before update × ***Interest Rate***
= ***Deposit Account*** (0) × 0.05
= 10,000 × 0.05
= 500
Therefore,
Deposit Account (1) = 10,000 + 500
= 10,500
Two years later, calculating in the same way,
Interest Income = 10,500 × 0.05
= 525
Deposit Account (2) = ***Deposit Account*** (1) + Interest Income
= 10,500 + 525
= 11,025

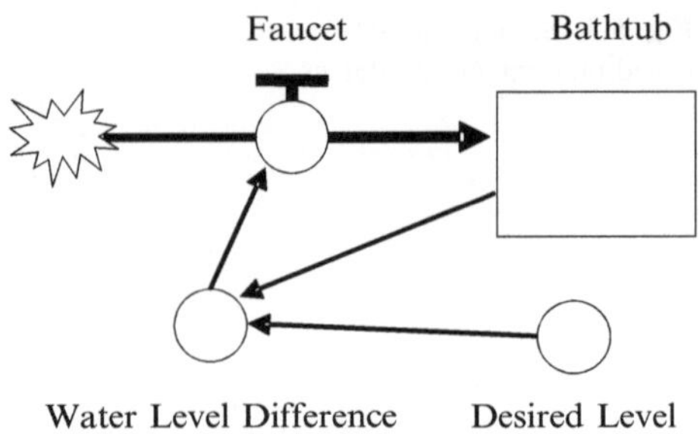

Fig. 5.4 Bathtub model with desired level

Thus, the increase in the value of the stock will cause the value of the inflow to rise. In turn, the value of the stock increases further. Such a relationship is called a positive feedback loop.

Next, we will explain a negative feedback loop using a Bathtub model with a desired level, as shown in Fig. 5.4. We set the value of the converter Desired Level to 100 l and the initial value of the stock Bathtub to 0 l. We redefine the converter Water Level Difference to "Water Level Difference = Desired Level – Bathtub" and the flow Faucet to "Faucet = Water Level Difference × 0.4." In this model, the difference between Desired Level and the water volume in Bathtub is calculated and 40% of this value is added to the stock Bathtub through the flow Faucet.

When we set the time interval (Δt) to 1 min and run this model, the change of stock Bathtub up to 2 min later is as follows:

After 1 min,
Bathtub (1) = ***Bathtub*** (0) + ***Faucet***
Water Level Difference = ***Desired Level*** – ***Bathtub*** (0)
= 100 – 0
= 100
Faucet = ***Water Level Difference*** × 0.4
= 100 × 0.4
= 40
Therefore,
Bathtub (1) = 0 + 40
= 40
Two minutes later, calculating in the same way,
Water Level Difference = 100 – 40
= 60
Faucet = 60 × 0.4
= 24
Bathtub (2) = ***Bathtub*** (1) + ***Faucet***
= 40 + 24
= 64

The behavior of the model in Fig. 5.4 is different from that of the model in Fig. 5.3. The water level difference gets smaller every minute. As a result, the water level in the bathtub gradually approaches the desired level because the increase rate of the stock is reduced. Such a feedback loop is called a negative feedback loop.

The simple models in Figs. 5.3 and 5.4 are examples of monotone changes. In some cases, a change in the opposite direction is only brought about under a special condition. If such reactions occur, the behavior of the system as a whole gradually converges to an equilibrium, but the trend sometimes exhibits cyclic changes or partial oscillatory behaviors. If multiple stocks, flows, and converters are connected by connectors, multiple feedback loops can occur in a model. In such a model, a loop starts from a given stock, goes through some converters of flows, and eventually returns to the starting stock. The behaviors of stocks or flows sometimes have an S-shape or create cyclic changes according to the combination of positive and negative feedback loops.

5.4 Example: A Local Development Model that Makes Efficient Use of a Rich Natural Environment

Now, let's make a model using the scenario shown below and run it in SD.

5.4.1 Scenario

Assume that agriculture and tourism are the main local industries in target region *A*, which makes efficient use of its rich natural environment. Tourists visiting region *A* look forward to the various activities available in this rich natural environment. Tourism contributes significantly to the local development of region *A*. In addition, resort development is also carried out to attract tourists and promote the vitality of the local economy. However, over-development results in environmental destruction. The higher the number of visiting tourists, the more strain the natural environment has to sustain. The more a rich natural environment is damaged, the less attractive it is for tourists, and the fewer tourists will visit. As a result, the region's vitality will decrease. We will now consider the relationships among Rich Natural Environment, Tourist Attractiveness, and Regional Vitality using SD.

First, we consider a rough causal loop diagram based on the above scenario. In addition to Rich Natural Environment, Tourist Attractiveness, and Regional Vitality, we include two converters: Sightseeing Load and Resort Development. The former represents the environmental load induced by visiting tourists. The latter corresponds to the positive effect of resort development on tourism. For these elements, we set the following assumptions:

- Increases in Regional Vitality stimulate efforts to preserve the natural environment.
- Increases in Tourist Attractiveness result in increased Regional Vitality.
- Increases in Rich Natural Environment result in increased Tourist Attractiveness.

The causal loop diagram based on these assumptions is shown in Fig. 5.5. If two elements are linked by a double-headed arrow, they are related to each other.

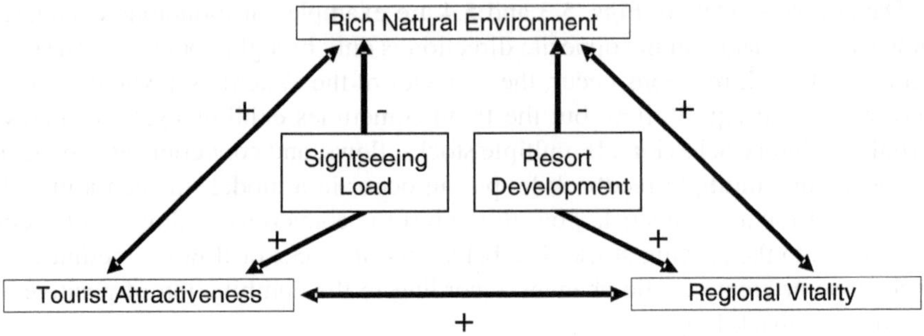

Fig. 5.5 Causal loop diagram

A single-headed arrow connecting two elements indicates a unidirectional relationship. A plus sign (+) beside an arrow means that the two elements have a positive feedback (both elements change in the same direction), while a minus sign (−) means that the two elements have a negative feedback (changes in these elements are in opposite directions). For example, increases in Rich Natural Environment result in increased Tourist Attractiveness, which in turn promotes an increase in Regional Vitality. The increase in Regional Vitality then induces an increase in Rich Natural Environment. On the other hand, since both Sightseeing Load and Resort Development have negative feedback loops to Rich Natural Environment, increases in these two factors result in decreases in Rich Natural Environment. Thus, Rich Natural Environment, Tourist Attractiveness, and Regional Vitality do not always change in the same direction because of the inclusion of negative feedback loops in the model.

5.4.2 Basic Model Based on a Scenario

Now, let us construct an SD model using the causal loop diagram shown in Fig. 5.5. First, we must determine stocks, flows, and converters. We define Rich Natural Environment, Tourist Attractiveness, and Regional Vitality as stocks because these elements change along a time scale. Sightseeing Load and Resort Development are assumed to be convertors because these elements are constants impacting the state of stocks. The SD model based on Fig. 5.5 is shown in Fig. 5.6.

The content of each stock, flow, and convertor in the model shown in Fig. 5.6 must be defined before running the simulation using SD. Since Rich Natural Environment, Tourist Attractiveness, and Regional Vitality are abstract concepts, we assign 50 units as the initial values for these stocks. In this context, unit is a fictitious unit system. If one of the factors (i.e., Rich Natural Environment, Tourist Attractiveness, or Regional Vitality) reaches over 50 units, it is considered to be in a

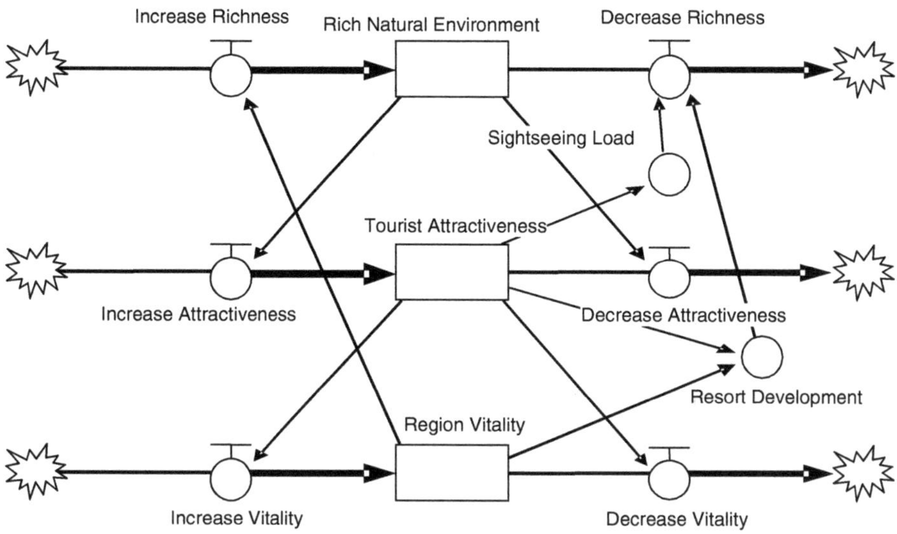

Fig. 5.6 SD model

desirable state, whereas it is considered to be in an undesirable state up to that point. The magnitude of this fictitious unit system has a mathematical meaning. For example, if the value of the stock Rich Natural Environment increases from 50 to 60 units, we can consider that Rich Natural Environment has gained 10 units.

The inflow to Rich Natural Environment (Increased Richness) is defined as depending on the state of Regional Vitality. We assume that if Regional Vitality is greater than 50 units, people are more interested in the preservation of the natural environment and it contributes 10 units to the increase in Rich Natural Environment. In contrast, if Regional Vitality is less than 50 units, people cannot make efforts to preserve the natural environment. We assume that the outflow from Rich Natural Environment (Decreased Richness) is determined by two convertors, namely Sightseeing Load and Resort Development. That is, if the sum of Sightseeing Load and Resort Development becomes positive, the natural environment is negatively impacted. Consequently, we define Rich Natural Environment as having decreased by 15 units. We assume that both the inflow to Tourist Attractiveness (Increased Attractiveness) and the outflow from it (Decreased Attractiveness) depend on Rich Natural Environment. Tourists are interested in the rich natural environment in region *A*, so Rich Natural Environment directly affects Tourist Attractiveness. Therefore, if the state of Rich Natural Environment is greater than 50 units, the value of the flow Increased Attractiveness will be 10 units, while if it is lower than 50 units, the value of the flow Decreased Attraction will be 15 units. Both the inflow to Regional Vitality (Increased Vitality) and the outflow from it (Deceased Vitality) are assumed to depend on the state of Tourist Attractiveness. If Tourist Attractiveness is higher than 50 units, the value of the flow Increased Vitality will be 15 units, while if it is lower than 50 units, the value of the flow Decreased Vitality will be 15 units.

The convertor Resort Development is set to 1 unit if both Tourist Attractiveness and Regional Vitality are greater than 50 units. We assume that the motivation to develop resorts is not enhanced simply if a large number of tourists visit the region (i.e., tourist attractiveness is high); enhancement also requires that the vitality of the region is sufficient. On the other hand, the convertor Sightseeing Load is set to 1 unit if Tourist Attractiveness is higher than 50 units.

Equations describing the model defined above are shown in Fig. 5.7. These equations cannot be input directly into STELLA. These equations are set by using the Model mode in STELLA. In this mode, we can click on any stock, flow, and convertor to set its equation using an input dialog displayed in STELLA.

We can run the simulation after setting the equations for the model. The model will run for 20 years. Stocks will be updated every year (i.e., the value of t is $0 \leq t \leq 20$ and $\Delta t = 1$ in the equations described in Fig. 5.7). The changes in the three stocks (i.e., Rich Natural Environment, Tourist Attractiveness, and Regional Vitality) are shown by the graph in Fig. 5.8. During the first year, since both Tourist Attractiveness (■) and Regional Vitality (▲) are equal to 50 units, which is their initial value (i.e., value at year 0), both Resort Development and Sightseeing Load are equal to 1 unit. As a consequence, Increased Richness is lower than Decreased Richness, resulting in a 5 unit decrease in Rich Natural Environment. During this period, both Tourist Attractiveness and Regional Vitality increase. While both the inflow and outflow of Regional Vitality are equal to 15 units, the inflow of Tourist Attractiveness is equal to 10 units and its outflow is 15 units. Tourist Attractiveness is more sensitive than Regional Vitality to the decrease in Rich Natural Environment, which is demonstrated by its decrease rate being greater than its increase rate. This is because both rates are directly related to Rich Natural Environment, unlike the rates dictating the evolution of Regional Vitality. As a result, the decrease in Tourist Attractiveness occurs earlier than the decrease in Regional Vitality. The increases and decreases in Regional Vitality are directly affected by Tourist Attractiveness alone, and only indirectly by increases and decreases in Rich Natural Environment, the impact of which is mediated by Tourist Attractiveness. In this model, the delayed responses of Tourist Attractiveness and Regional Vitality are induced by the increases and decreases in Rich Natural Environment. Thus, every stock repeatedly rises and falls during the 20 modeled years.

The expression "Increased Richness = if (Regional Vitality ≥50) then (10) else (0)" is a so-called *if-statement*. The format of an if-statement is "if (condition) then (statement or value) else (statement or value)." If the condition is true, the statement or value in the *then* block is selected. In contrast, if the condition is false, the *else* block is selected. For example, a statement "Increased Richness = if (Regional Vitality ≥50) then (10) else (0)" means that Increased Richness is assigned 10 units if Regional Vitality is greater than 50 units, while it is assigned 0 units if Regional Vitality is lower than 50 units. In the case when multiple conditions are used, such as "Resort Development = if [(Tourist Attractiveness ≥50) AND (Regional Vitality ≥50)] then (1) else (0)," operators such as *AND* or *OR* are used. A statement

Rich Natural Environment (t) =
Rich Natural Environment (t $-\Delta t$) +
(Increase Richness – Decrease Richness) $\times \Delta t$
Initial value: Rich Natural Environment = 50
Inflow:

Increase Richness = if (Regional Vitality >= 50)
then (10) else (0)

Outflow:

Decrease Richness = if ((Resort Development +
Sightseeing Load) > 0)
then (15) else (0)

Tourist Attractiveness (t) = Tourist Attractiveness (t $-\Delta t$) +
(Increased Attractiveness –
Decreased Attractiveness) $\times \Delta t$
Initial value: Tourist Attractiveness = 50
Inflow:

Increased Attractiveness =
if (Rich Natural Environment > = 50)
then (10) else (0)

Outflow:

Decreased Attractiveness =
if (Rich Natural Environment < 50)
then (15) else (0)

Regional Vitality (t) = Regional Vitality (t $-\Delta t$) +
(Increased Vitality – Decreased Vitality) $\times \Delta t$
Initial value = 50
Inflow:

Increased Vitality = if (Tourist Attractiveness > =50)
then (15) else (0)

Outflow:

Decreased Vitality = if (Tourist Attractiveness < 50)
then (15) else (0)

Resort Development = if ((Tourist Attractiveness >= 50) AND
(Regional Vitality >= 50)) then (1) else (0)

Sightseeing Load = if (Tourist Attractiveness >= 50)
then (1) else (0)

Fig. 5.7 Equations for the model in Fig. 5.6

"Resort Development = if [(Tourist Attractiveness $\geq$50) AND (Regional Vitality $\geq$50)] then (1) else (0)" means that Resort Development is assigned 1 unit if and only if both Tourist Attractiveness and Regional Vitality are higher than 50 units; otherwise it is assigned 0 units.

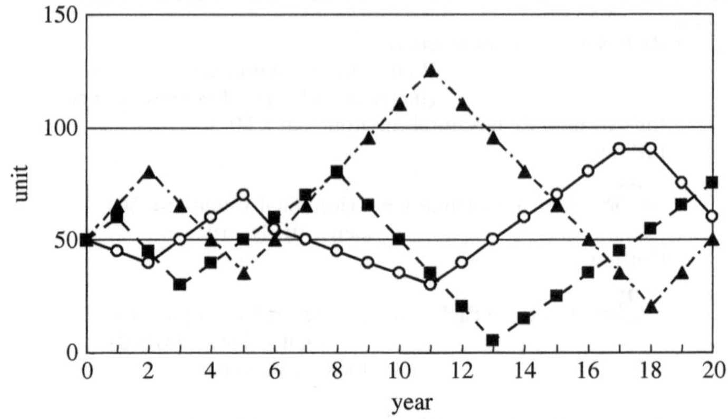

Fig. 5.8 Changes in each stock. ○: Rich natural environment, ■: Tourist attractiveness, ▲: Regional vitality

5.4.3 Comparison Between Scenarios Using SD

In the model shown in Fig. 5.6, we assume that Regional Vitality and Resort Development are the factors affecting Rich Natural Environment and that the flow Increased Richness is always set to 10 units when Regional Vitality is higher than 50 units. In addition, Resort Development only affects Decreased Richness and does not directly affect Increased Vitality. Results from SD simulations under these assumptions have demonstrated that changes in Rich Natural Environment affect Tourist Attractiveness and Regional Vitality. Next, let's examine the kinds of considerations that are possible using SD if we wish to decide the best balance between environmental preservation and resort development. A simple example is shown as follows.

Assume that we will make some efforts to preserve the natural environment to maintain and improve Rich Natural Environment. Although resort development is one of the causes of environmental destruction, it is also expected to help local development by attracting tourists and promoting employment in the region. Therefore, we modify the model in Fig. 5.6 by considering efforts to preserve the natural environment and the effects of resort development. The new model is presented in Fig. 5.9.

In the new model, we assume that the convertors Preservation Coefficient and Development Coefficient are linked to Preservation Effort and Resort Development, respectively. These convertors stand for constants (coefficients). When executing simulations in which various parameter values change, it is useful to define the target parameters as independent convertors. Although not described here, these kinds of convertors can be handled with Sliders or Knobs in the interface layer of STELLA. Equations corresponding to Fig. 5.9 are shown in Fig. 5.10.

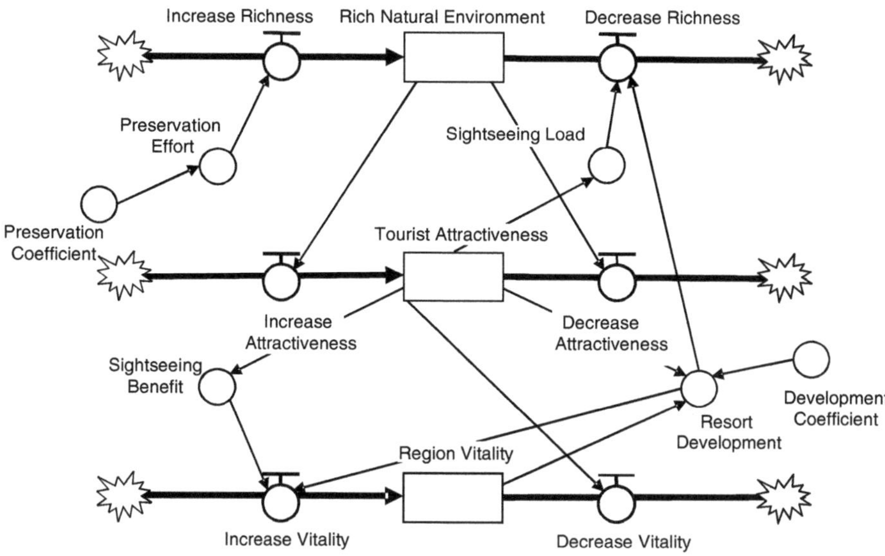

Fig. 5.9 Model considering environmental preservations efforts and the effects of resort development

The evolution of each stock under different combinations of changes in the Preservation Coefficient and Development Coefficient are shown in Fig. 5.11. We consider the following four scenarios:

(a) No efforts for environment preservation (Preservation Coefficient = 0 units) but high resort development (Development Coefficient = 5 units).
(b) Preservation Coefficient is set to 5 units but no resort development (Development Coefficient = 0 units).
(c) Both Preservation Coefficient and Development Coefficient are set to 5 units.
(d) Preservation Coefficient is set to 15 units and Development Coefficient is set to 5 units.

The results of the scenario with resort development and no environmental preservation efforts show that Rich Natural Environment becomes nonexistent (i.e., 0 unit) in the fifth year, and both Tourist Attractiveness and Regional Vitality temporarily increase but decrease rapidly thereafter (Fig. 5.11a). In the case where both Preservation Coefficient and Development Coefficient were set to 5 units (Fig. 5.11c), the changes in Tourist Attractiveness and Regional Vitality showed the same behavior as in Fig. 5.11a. However, Rich Natural Environment exhibited only a slight change and recovered its initial value (50 units) in year 13, thanks to the Preservation Coefficient being set to 5 units/year in the model run shown in Fig. 5.11c. On the other hand, all stocks showed cyclic behaviors when the Preservation Coefficient was set to 5 units and the Development Coefficient was set to 0 units (Fig. 5.11b). Regional Vitality decreased gradually with repeated increases and decreases. When we put a large weight on environmental preservation

Rich Natural Environment (t) =
Rich Natural Environment (t – Δt) +
(Increased Richness – Decreased Richness) × Δt
Initial Value: Rich Natural Environment = 50
Inflow:
Increased Richness = Preservation Effort
Outflow:
Sightseeing Load + Resort Development

Tourist Attractiveness (t) = Tourist Attractiveness (t –Δt) +
(Increased Attractiveness –
Decreased Attractiveness) × Δt
Initial value: Tourist Attractiveness = 50
Inflow:
Increased Attractiveness =
if (Rich Natural Environment > = 50)
then (10) else (0)
Outflow:
Decreased Attractiveness =
if (Rich Natural Environment < 50)
then (15) else (0)

Regional Vitality (t) = Regional Vitality (t –Δt) +
(Increased Vitality – Decreased Vitality) × Δt
Initial Value: Regional Vitality = 50
Inflow:
Increased Vitality = Sightseeing Benefit +
Resort Development
Outflow:
Decreased Vitality = if (Tourist Attractiveness < 50)
then (10) else (0)

Resort Development = if ((Tourist Attractiveness > 40)
then (Development Coefficient) else (0)
Sightseeing Load = if (Tourist Attractiveness > = 50)
then (5) else (0)
Preservation Effort = if (Rich Natural Environment < 50)
then (Preservation Coefficient) else (0)
Sightseeing Benefit = if (Tourist Attractiveness >= 50)
then (10) else (0)
Development Coefficient = 5
Preservation Coefficient = 5

Fig. 5.10 Equations for the model presented in Fig. 5.9

(Preservation Coefficient = 15 units, Development Coefficient = 5 units), all stocks showed the same cyclic behaviors as in Fig. 5.11b, but Regional Vitality was always greater than its initial value (50 units). Both Rich Natural Environment and Tourist Attractiveness oscillated around their initial value (50 units) within a range of ±10 units. Comparison of the four results shown in Fig. 5.11 indicates that

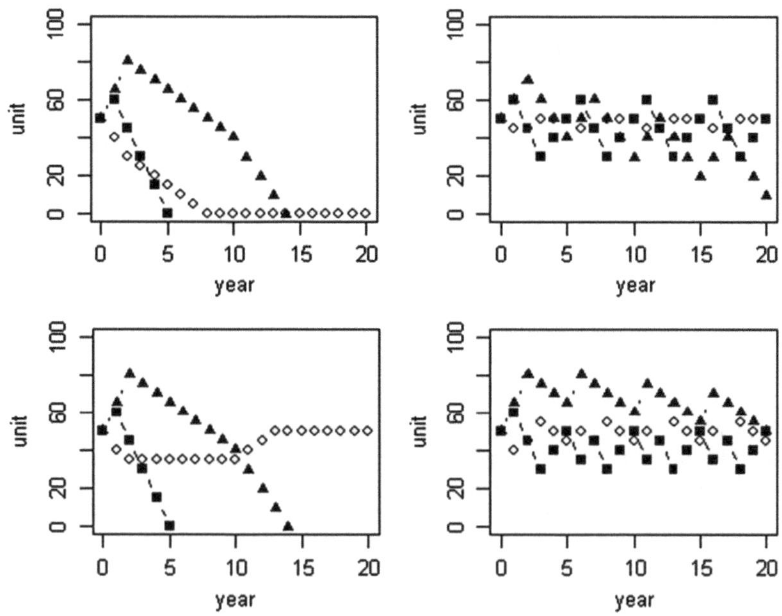

Fig. 5.11 Evolution of each stock. Legend ○: Rich natural environment, ■: Tourist attractiveness, ▲: Regional vitality. *Top left* (**a**): Preservation coefficient = 0 units, Development coefficient = 5 units. *Top right* (**b**): Preservation coefficient = 5 units, Development coefficient = 0 units. *Bottom left* (**c**): Preservation coefficient = 5 units, Development coefficient = 5 units. *Bottom right* (**d**): Preservation coefficient = 15 units, Development coefficient = 5 units

it is important to preserve the natural environment but it is also necessary to develop resorts to a certain extent to promote regional development based on a rich natural environment.

5.5 Further Improvements

In this chapter, we considered models with very simple assumptions as examples of simulations using SD. Although the models shown here are not sufficient to solve problems in actual societal systems, they can be helpful for understanding the concept of SD and how to construct models. If we wish to expand these models into societal models for regional development, the following points must be taken into account.

First, we must define a more precise method for evaluating the value of Rich Natural Environment. For example, we can consider the size of areas to be preserved or the proper method of preservation by classifying all land use in the region using remote sensing and by evaluating changes in the natural environment up to the present using remotely sensed time series data. Sightseeing load and the effects of resort development can also be considered more accurately using an advanced ecosystem model. With regard to Tourist Attractiveness, it is useful to consider an evaluation index based on dynamic statistics of tourist numbers or to conduct a questionnaire survey. For Regional Vitality, we can develop more detailed models by incorporating information such as the industrial structure of the region, the age structure of employees, the movement of the population, and the economic index in addition to information concerning resort development and tourism.

References

Forrester, J.W. *Industrial Dynamics.*, Pegasus Communication, 1961

Forrester, J.W. *Urban Dynamics.*, Pegasus Communication, 1966

Meadows, D.H., Meadows, D.L., Randers, J. *Beyond the Limits.* Chelsea Green Pub., 1992

Meadows, D.H., Randers, J, Meadows, D.L. *Limits to Growth – The 30-Year Update*, Chelsea Green Pub., 2004

Liu, C., Yoshimura, T., Moriya, K., Sakai, T. *A simulation of beef cattle production integrated with rice cultivation using system dynamics (in Japanese).* Bulletin of Beef Cattle Science 72 59–66, 2002

Deaton, M.L., Winebrake, J.I. *Dynamic Modeling of Environmental Systems.* Springer, 1999

Chapter 6
Multiagent Simulation

Toru Ishida, Hiromitsu Hattori, and Yuu Nakajima

Abstract How can we predict ICT-driven innovations that will emerge in society and daily life? Multiagent simulations can be used to predict the changes in society and daily life caused by human interaction with new technologies. Multiagent simulations have become increasingly popular as a type of micro-simulation that can represent diversity and heterogeneity of human behaviors. In contrast to traditional micro-simulations, multiagent simulations can represent individual decision-making in detail, so they can reproduce the complex phenomena that arise from the outcome of interactions between different agents. In this chapter, we introduce three cases where multiagent simulations are used to reproduce and analyze complex collective behavior; evacuation simulation, traffic simulation, and economic simulation. Furthermore, we discuss the participatory approach for realizing practical and reliable multiagent simulations.

6.1 Why Multiagent Simulations?

The reason information technology is thought to have had such a large impact on society is that it accelerates the fragmentation and recombination of various social systems. For example, since information technology has made communication more efficient and largely reduced organizational costs, various communities have evolved. As a result, people do not belong to a single community but participate in multiple different communities. Also, because information technology has made data transfer and data sharing easier, the result is a demand for increased independence of elements within organizations, and for mutual interaction and cooperation on a wide scope. A variety of phenomena that have been

---

T. Ishida (✉) • H. Hattori • Y. Nakajima
Graduate School of Informatics, Kyoto University, Yoshida-Honmachi, Sakyo-ku, Kyoto 606-8501, Japan
e-mail: ishida@i.kyoto-u.ac.jp; hatto@i.kyoto-u.ac.jp; nkjm@i.kyoto-u.ac.jp

T. Ishida (ed.), *Field Informatics*, DOI 10.1007/978-3-642-29006-0_6,
© Springer-Verlag Berlin Heidelberg 2012

taken for granted, such as hierarchical organizational structures, or taken as common sense, such as social norms, are being reexamined and many changes are emerging.

Innovations emerge with the appearance of new technologies as well as by the elicitation of needs in society and daily life. ICT technologies, which can facilitate fragmentation and recombination of social systems, boost such innovations and inspire society with belief of unlimited possibilities. The question is how to predict ICT-driven innovations emerging in society and daily life? The required simulations are not traditional ones, such as physical or chemical simulations, but rather are novel simulations that can calculate a sequence of interactions between humans/organizations based on individual models of autonomous decision-making entities. Multiagent simulations introduced in this chapter can be used to predict possible changes in society and daily life triggered by human needs and interaction with new technologies.

Essentially, simulations are used to virtually reproduce complex phenomena that are difficult to observe in the real world, on computers. Simulations can be divided into two classes; macroscopic and microscopic simulations according to the abstraction level of models of the simulation targets. System dynamics is a typical example of a macroscopic simulation. Macro simulation reproduces a phenomenon based on a macroscopic viewpoint in that the entire simulation target is represented as a single model and its behavior is defined by governing equations. Consequently, macro simulations allow the observation of behaviors or changes in the overall system, but the local properties of individual elements or interactions among elements are not reproduced. On the other hand, micro simulations reproduce a complex social phenomenon by accumulating the microscopic behaviors of models of social entities (e.g., humans or organizations) including the interaction among them. Assuming that human society consists of a lot of decision-making entities, it seems natural to use micro simulations to predict the behavior of society. In particular, micro simulations have manifested their ability to clearly present a variety of individual core behaviors in the reproduction and analysis of the following kinds of complex collective behavior.

6.1.1 Evacuation Simulation

In a disaster evacuation simulation, the evacuees and rescue workers have different attributes given the disaster conditions (e.g., collapsed houses, fire, etc.) and it simulates the conditions under which they perform evacuation behaviors and rescue activities, respectively. At such a time, the behaviors of the evacuees and the rescue workers are not the same, so it is necessary to analyze their behaviors individually. Furthermore, an analysis of micro–macro relationship is needed because panic may be fueled by a chain-reaction of local behaviors.

6.1.2 *Traffic Simulation*

Traffic is a social phenomenon that is represented as an accumulation of interactions among vehicles caused by drivers' behaviors. Drivers will act in a variety of ways even under identical traffic conditions due to their individuality, cognitive ability, as well as the performance of the vehicles they drive. This results in a wide variety of traffic phenomena. Therefore, to efficiently control traffic systems for ensuring smooth traffic flow, it is desirable to comprehend the changes that emerge in the overall traffic pattern created by each driver's behavior and the setting of traffic controls such as speed limits. In other words, if we could run a simulation in advance that applied a variety of traffic policies and then compared the effects and influence of each policy, we would be able to more effectively optimize the policy mechanisms, such as traffic rules.

6.1.3 *Economic Simulation*

Market economy analysis based on simulations has a long history. Even today researchers are very active in looking at the construction of artificial markets and the reproduction of economic phenomena. The market is a social system where participants have various and complicated preferences and strategies. Traditional economic theories usually assume the rigid (sometimes unrealistic) rationality of players in the market, but irrational behaviors can be observed in the real market. What is required then is to combine theoretical analysis with its assumption of rational players with behavioral analysis of real-life players and its more practical viewpoint. With economic simulations it is desirable to model and analyze the economic behavior of each player under a certain market mechanism.

Multiagent simulations have become increasingly popular as a type of micro simulation since they can well represent the diversity and heterogeneity of the behaviors of entities and enable us to observe micro and macro phenomena. In multiagent simulations, humans and organizations are modeled as agents. An agent can autonomously determine his/her behavior while interacting with the environment and other agents based on the surrounding environment and information from other agents. In contrast to traditional micro-simulations, which tend to enforce uniform entity behavior mechanisms, multiagent simulations can represent individual decision-making in detail according to each agent's circumstances, so they can reproduce the complex phenomena that arise from the results of interactions between different agents.

6.2 Creating and Analyzing Social Systems

Because multiagent simulations can straightforwardly represent the actions of individual people and organizations, they are being applied to social, economic and cultural problems which are rooted in human decision-making and have been difficult to experiment with. Simulations that represent complex social and economic problems are called social simulations (Gilbert and Troitzsch 1999) and the application of multiagent simulations in social simulations can be broadly broken down into the following two approaches.

6.2.1 Social System Analysis

Simulations are used to understand social phenomena and to analyze social systems. Simulations are also used as a method for conducting experiments or to prove hypotheses in the social sciences. Multiagent simulations are used in order to understand the relationships between individual attributes and behavior (micro level) and social groups (macro level) in emergent behavior (phenomena on completely different levels are generated from interaction between factors). That said, when using an analysis type of simulation, if the agent model becomes complex, it becomes difficult to understand the causal relationships between the attribute values of the model and the results of the simulation. Thus, the model of each agent is kept simple and complexity arises from the interaction between agents (Axelrod 1997); simulations are normally modeled according to the KISS principle (Keep It Simple, Stupid).

6.2.2 Social System Synthesis

Social simulations are also used in the design of new social systems and in the experimentation and training in the initial stages of developing complex systems. Such synthesis types of simulations put emphasis on systems and the development and verification of systems, so they do not concern themselves with the complexity of agent modeling. For example, in the Robocop Rescue project, a disaster rescue simulator is being developed to anticipate what will happen when earthquakes or fires occur in urban centers (Tadokoro and Kitano 2000). This simulator is being used to simulate how humans or robots would perform rescue efforts in actual environments, and the simulation results can be used to develop a new rescue support system.

The synthesis type of simulation is also used to learn how different people would utilize socially embedded information systems, such as ubiquitous computing. Simulations are very useful since many experiments are difficult or impossible to conduct in the real world. When a simulation aims to synthesize a social system, researchers can also arrange for people to participate in the simulation in order to reflect the various points of views of the stakeholders (participatory simulation). As will be discussed later, multiagent simulations can easily be extended to participatory simulations.

Agent modeling is required for both of these approaches, analysis type and synthesis type, but the degree of detail required in each differs. In the analysis type, patterns of behavior in the problem area are categorized, sorted and a model is implemented based on each pattern. For example, for a simulation that analyzes a market in which there are multiple buyers and sellers, researchers analyze patterns of trading strategies based on accumulated data, the literature and specialist knowledge, and implement the patterns as agents (Izumi 2003). Conversely, in a synthesis type of simulation, the need arises to grasp phenomena in more detail (fine grain) than in the analysis type. For example, a simulation that seeks to prove the validity of a newly designed market has to model the trading strategies that reflect the characters and peculiarities of individual buyers and sellers in a real world market.

Following is a discussion of the structure and illustrations of multiagent simulations that seek to synthesize new social systems.

6.3 Participatory Simulations

Because multiagent simulations are conducted at the human or organizational level, they have a high degree of affinity to participatory simulations in which people participate in the simulation process. Participatory simulations can be broadly divided into the following two kinds of forms.

In the first form, a simulation is used for training that takes place in a virtual environment or the simulation is used to experiment with a new system. Accordingly, no distinction is made between humans and agents and interactions take place among them and/or with the environment. For example, in a participatory disaster drill system, the simulation proceeds with trained participants and groups of agents fulfilling their roles, such as evacuation leaders or evacuees. In contrast with training systems to date, such as flight simulators, in which the main aim is to learn how to manipulate a machine, in a participatory multiagent simulation, participants learn skills such as how to assess a situation and how to communicate in real life, such as when there is a disaster or on a battlefield. Furthermore, if a three-dimensional virtual space is incorporated, it is possible to create a place to make realistic decisions by subjecting the user to a virtual experience; by observing and analyzing the behaviors of subjects in the virtual space, the agent model can be refined and a more accurate simulation can be achieved.

The second form uses simulations for mechanism design. In this case, reflecting stakeholder viewpoints in the simulation yields realistic results and the accumulation of experience. Examples of this type are having residents address issues such as a community building or where to place a garbage processing plant. When stakeholders take part in the problem solving process, a realistic solution is more likely to be obtained than when the bureaucrats propose a top-down solution. Accordingly, we modeled people's decision-making by having stakeholders participate in gaming session, which we visualized as a simulation; it provides a place in which the model can be verified. In doing so, we adopted the diverse viewpoints of stakeholders into the mechanism design.

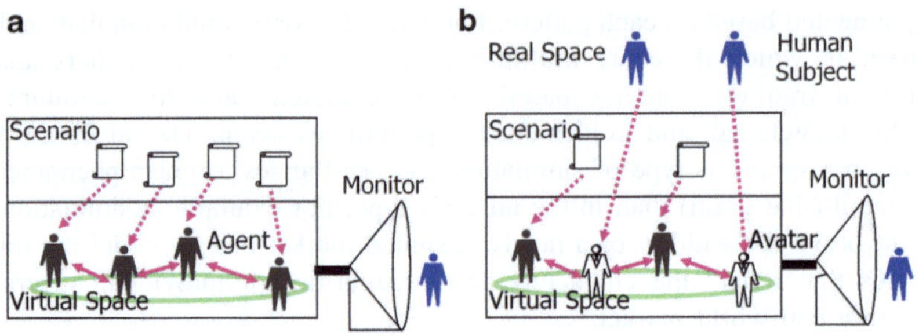

Fig. 6.1 System architecture of simulators: (**a**) multiagent simulation, (**b**) multiagent participatory simulation

The system architecture of a participatory simulator is an expanded version of that of a multiagent simulator. Figure 6.1a illustrates the system architecture of a multiagent simulator. The behavior of each and every agent is controlled by the description of the simulation scenario. Next, this is extended to the participatory simulator illustrated in Fig. 6.1b. In other words, some of the scenario-controlled agents are replaced by human-controlled avatars. As a result, a situation is created in which scenario-controlled and human-controlled agents mix together in the virtual space. Thus, although the agents and avatars are controlled by different methods, the simulation environment does not discriminate between them. To acquire the simulation results, we monitor the entire process of the simulation by observing the virtual space.

To gain realistic knowledge from the results of a multiagent simulation, there cannot be a huge gap between the simulation environment and the real world, and the simulation has to be such that people can participate without feeling uncomfortable. It is worth pointing out that once the information on human behaviors is abstracted and their details are eliminated, there could be lack of information for constructing a realistic agent behavior model. Further, it is extremely difficult to extrapolate individual human behaviors from social statistical data alone. Thus, extracting human behaviors within a problem domain requires techniques for modeling them.

The architecture of participatory simulations is illustrated in Fig. 6.1b. Participatory simulations provide a means for observing human behaviors in a problem domain. That is to say, participator simulations implement three processes (1) runs the simulation, (2) constructs an agent model from simulation results (or refines the model), and (3) designs and verifies the mechanisms based on the simulation.

In summary, as in Fig. 6.1, a participatory simulation consists of (1) agents for modeling users or organizations, (2) scenarios for modeling interactions, (3) virtual space to represent real space, (4) a monitor to visualize simulations ongoing in the virtual space, (5) avatars to represent human subjects and (6) human subjects to control avatars.

6.4 Examples of Multiagent Simulations

6.4.1 Evacuation Simulations

Following is a description of examples of multiagent simulations using the virtual city simulator FreeWalk/Q (Nakanishi and Ishida 2007), which was developed by the Graduate School of Informatics, Kyoto University. The simulator is structured as a virtual city space in which verbal communication among multiple users is possible. Multiple users can gather via the network, and use their avatars to represent themselves. Avatars close to one another in the virtual city space can carry on verbal conversations. In addition, the simulator features a function that displays a 3D model of the city, a walking animation function to increase the reality of avatars, a model for avoiding collisions in physical interactions, as well as a gesture function for non-verbal communication. With just these functions, evacuation drills conducted at schools and workplaces can be conducted in virtual space.

We introduce scenario description languages to describe agent's decision-making in the virtual city space. Although a scenario description language capable of describing agent's internal states, such as their beliefs, intentions, etc., is a high-level matter, a much simpler thing is a state transition model, which describes the behavior choices according to the agent's situation. Figure 6.2 indicates an example of a state transition model as written in the scenario description language Q. Question mark "?" represents agent's perception (called cue), and exclamation mark "!" represents agent's action (called action). For example, Fig. 6.2 is part of the leader's scenario. The first four lines mean "if the leader sees an evacuee standing in State1, he/she tells the evacuee to follow him/her and transits to State2." In the simulation, behavior of leaders and evacuees are described using such statetransition diagrams.

The process of conducting a simulation is as below. First, a scenario writer (professionals in the application domain, here, evacuation domain) and an agent system developer (professionals for multiagent systems) agree upon cues and actions as the interface between them. Second, the scenario writer describes scenarios using the determined cues and actions, while the agent system developer implements the cues and actions in the virtual city simulator. An additional benefit yielded by the introduction of the scenario description language is that a clear interface can be specified between scenario writers and agent system developers.

Using the virtual city simulator and scenario description language Q, the controlled experiment conducted by Prof. Toshio Sugiman was simulated. The purpose of this experiment was to compare two evacuation methods; the "Follow-direction method" and the "Follow-me method." In the "Follow-direction method," the leader shouts out, "Exit is over there. Go over there," while indicating the direction of the exit, and eventually moves toward the exit. This method is the conventional approach in evacuation drill. On the other hand, in the "Follow-me method," the leader tells a few of the nearest evacuees to follow him/her and actually proceeds to the exit without verbalizing the direction of the exit.

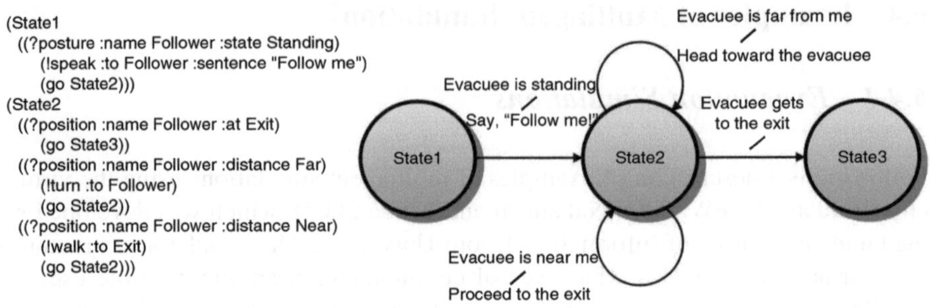

Fig. 6.2 Example of a state transition model

Let us reproduce the experiment that Prof. Sugiman conducted in 1988 in a virtual city space. First, we describe scenarios for leaders employing the "Follow-direction method," leaders employing the "Follow-me method" and evacuees. Evacuees do not simply follow the instructions of leaders, but follow to some extent the behavior of the group around them. The possible scenarios are described below.

1. Leader's scenario ("Follow-direction method")

When the leader sees evacuees, he/she indicates the direction of the exit in a loud voice. When the leader spots an evacuee heading towards the wrong exit, he/she shouts out a warning. When the leader finds that all the evacuees around him/her are correctly evacuating, he/she joins them. When the leader finds an evacuee who is not moving, he/she directly encourages him/her to move to the exit.

2. Leader's Scenario ("Follow-me method")

The leader tells a few of the nearest evacuees to follow him/her and actually proceeds to the exit without verbalizing the direction of the exit. When the leader notices that the evacuees following him/her are falling behind, he waits for them to catch up. After the evacuees have caught up, the leader resumes heading toward the evacuation point. If the leader misses the evacuees, he/she spots other evacuees by looking around. The leader repeats the above procedure until he/she gets to the exit.

3. Evacuee's Scenario

The evacuee follows the leaders' instruction. For the "Follow-direction method," the evacuee goes in the direction indicated by the leader. For the "Follow-me method," the evacuee follows the leader. If the evacuee cannot receive any instruction, he/she follows the crowd around him/her. If the evacuee sees an exit near him/her, he/she heads toward it.

Figure 6.3 shows a multiagent simulation. Figure 6.3a is a visualization of the virtual space from the agents' viewpoint, while Fig. 6.3b is a bird's-eye view of the same space. The former is called the immanent mode and the latter the transcendent mode. We refined the scenarios by analyzing the video of the experiment conducted in 1988 and interviewing Prof. Sugiman, and successfully double-checked the

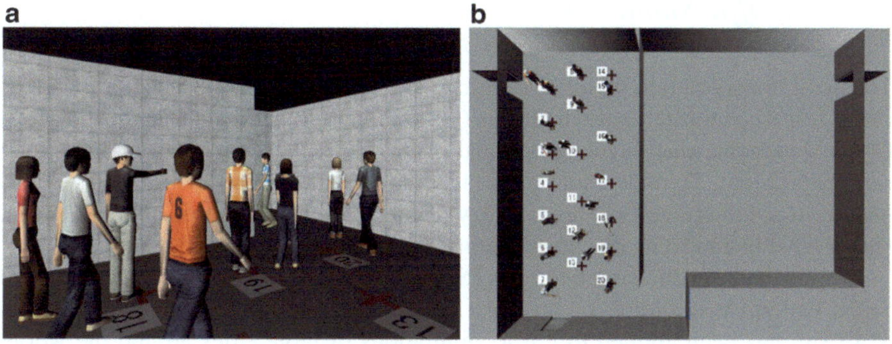

Fig. 6.3 Evacuation simulation (**a**) immanent mode, (**b**) transcendent mode

results of the previous controlled experiment conducted in a real environment. Moreover, we conducted participatory simulations wherein agents and avatars controlled by human subjects coexisted. When people actually took part as avatars, the simulation did not proceed as predicted. In contrast with the scenario-controlled evacuee agents, the human subjects exhibited diverse behaviors, many of which were unexpected.

Subsequently, using the data from the participatory simulation, we remodeled the evacuee behavior and fine tuned the evacuation method. The advantages of the modeling process used in the participatory simulation are described below. One of the advantages of participatory simulations in the modeling process is the possibility of extracting observations of each human subject and their operation history from log data. That is, we can acquire behavior rules that represent what actions each subject adopted in each environment. However, if we stop the modeling process at this point, we cannot determine a causal relation between environmental conditions and behavior. Thus, next we interviewed human subjects while showing them a video of the simulation to determine if it is possible to explain the behaviors of the human subjects at each moment with our extrapolated behavior rules. Upon acquiring the operation history of human subjects via participatory simulations, we refined the agent models represented by the set of behavior rules (Ishida et al. 2007).

We conducted a multiagent simulation using the acquired agent models, where we varied the number of leaders for 16 evacuees. As a result, the simulation exhibited a similar difference between the simulated "Follow-me method" and "Follow-direction method" as that seen in the real world experiment. That is, the "Follow-me method" is superior to the "Follow-direction method" if the number of leaders is enough. This result illustrates that in environments such as department stores or underground shopping centers, where many leaders can be deployed, the "Follow-me method" is appropriate. In this way, we can shed light on the potential and limitations of new evacuation methods, and show that it is possible to improve an evacuation drill system for real world application.

6.4.2 Traffic Simulations

Traffic or transportation involves serious social problems. For instance, traffic jams cause large economic losses in Japan (12 trillion JPY). In countries like the US, China, and India, that have a great deal of land, it is recognized as an even more serious problem. To understand traffic is to understand the behavior of the various people in a society and the impact it will have on the design of compact cities, which have begun to be discussed as model cities of the future.

We introduce traffic simulator MATSim (Raney et al. 2003) implemented by the Swiss Federal Institute of Technology Zurich and Technische Universität Berlin. MATSim has been applied to traffic simulations of large cities around the world. MATSim has the following features.

6.4.2.1 Multiagent Traffic Simulation

Dynamic traffic assignment (DTA) is a typical method for traditional traffic simulations. DTA divides a city into a number of zones, origin–destination (OD) points, which represent movement between zones, are assigned, and traffic features, such as average travel time, are output. In this approach, however, the different characteristic behaviors of each driver could not be reflected in the simulations. MATSim represents each driver as an agent. The simulation platform assigns parameters to each agent, for example, origin, destination, mode of transportation and lets each agent calculate an optimal route. To put it another way, previous traffic simulations calculated the network flow that met OD matrices, but MATSim calculates the routes that fulfill the OD of each agent and by repeating the simulation, it learns what the best routes are.

6.4.2.2 Double-Layered Simulation

MATSim simulations are conducted through the collaboration of two layers: physical layer and mental layer. Agents move on the road network on the physical layer and generate route and daily activity plan on the mental layer. The physical layer is made up of links (roads) and nodes (intersections) and agents implement their plans of action decided in their mental layers. It is possible by using just the physical layer to analyze phenomena such as traffic jams that emerge from the reactive behaviors of groups of cars or adaptive traffic signal schemes. However, it cannot cope with human intentions, such as where cars are headed, what a driver is trying to accomplish in moving a vehicle, and so on. Therefore, the mental layer is introduced to the system which allows each individual's activity plan and route choices to be incorporated into the simulation. After the simulation ends each day, the results are evaluated and route choices are revised based on a predetermined strategy. Through the iterative improvement, agents' route plans converge on Nash equilibrium.[1]

MATSim was applied to a traffic simulation for the whole of Switzerland (Raney et al. 2003). First, Switzerland was divided into a number of zones. Based on the table of ODs between the zones, ODs were created for each agent. The OD matrices were provided by a Swiss regional planning authority (Bundesamt für Raumentwicklung). Next, by repeating the following steps, we calculated the ideal route for each agent. First, the data necessary for the simulation, such as agent attributes, initial route, road networks, etc., were input into a database, and a traffic simulation was run based on this input. Once the simulation began, events were generated by the timing of cars stopping due to reaching their destination or traffic jams. Based on the record of generated events, an agent computes the score of his/her route plan. For example, the plan score can be defined as the summation of differences between the ideal travel time and simulated travel time. Previously selected routes and their scores are input for each agent and the process of revising the route outputs the route that should be taken in the next simulation.

The results of simulations on MATSim were evaluated by comparison against observed data. Furthermore, results of traditional DTA-based simulations were also evaluated in the same way. The result was that MATSim yielded an average margin of error of −5.26 %. This means that when compared with actual traffic conditions, MATSim underestimated the number of cars on the road by an average of about 5 %. By contrast, the DTA simulation had an average relative error of about 16 %. These results indicate that multiagent simulation is better at reproducing the traffic phenomena that emerge from the interaction among drivers, each of who is an independent decision-maker, than traditional simulation methods.

How to use a multiagent simulation when designing a new traffic/transportation system? It is essential to incorporate realistic driving behaviors into simulations for realizing traffic simulations that are useful for mechanism design (Hattori et al. 2011). As discussed above, participatory simulations provide a modeling method that can extract and reflect the characteristics and individuality of each driver's driving operation. The mechanism design process consists of three steps: (1) Observing human behavior in a driving simulation and acquiring a driving model from a driving simulation, (2) run a large-scale traffic simulation based on the acquired model, and (3) analyze the results of the simulation and improve the mechanisms.

Figure 6.4 illustrates this process. A real traffic environment is reproduced in a virtual space and driving simulations where agents with initial model and human subjects participate are conducted. Although it is difficult to conduct experiments on traffic in the real world, a traffic simulation enables repeated observations of how people drive under controlled traffic conditions and to obtain observed data. In parallel with the driving simulation, interviews with human subjects are conducted

[1] Here the Nash equilibrium is the situation in which each driver chooses the best possible route, within the constraint of independently deciding their own behavior.

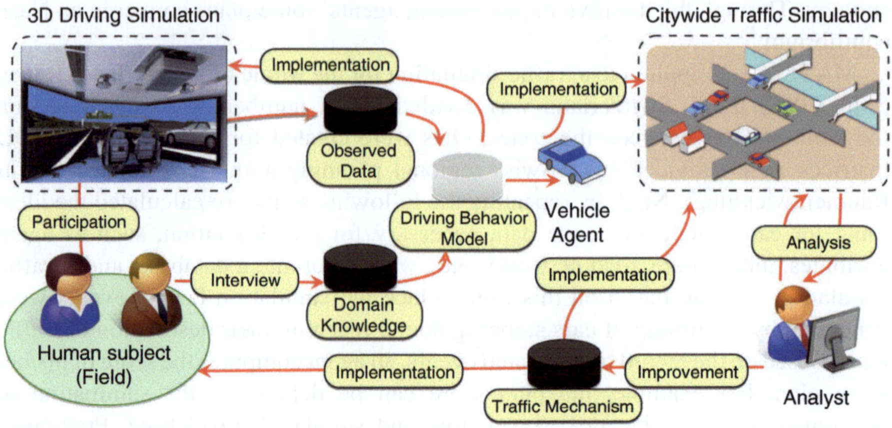

Fig. 6.4 Process of traffic simulation

to collect explanations about driving operations. Collected explanation is called domain knowledge. An example of such knowledge is "when I come to a hill, I know the car will slow down, so I step on the accelerator." Based on the observation data gathered in the driving simulation and domain knowledge, we obtain explanations of what the human subjects observed and how what they felt. The aggregate of the explanations is a driver model. The obtained models are used for reproducing the driving behaviors of people in traffic simulations. By improving the driver models via simulations in this way, driving behaviors, such as how vehicles speed up/slow down, change lanes, overtake others, etc., approaches that of humans, so as a result, the traffic phenomena that emerge from the simulations approach the real-life phenomena. Furthermore, through simulations under various traffic mechanisms using the driving behavior models obtained from human subjects, it becomes possible to estimate the effect of the introduction of traffic mechanisms. Thus, multiagent simulations enable us to analyze the reaction to new traffic mechanisms.

6.4.3 Economic Simulations

CIRAD (Centre de Coopération Internationale en Recherche Agronomique pour le Développement) uses participatory simulations for understanding farmers' land-use decisions in Vietnam and Thailand in a collaboration project of IRRI (International Rice Research Institute). The Graduate School of Informatics, Kyoto University, which developed the participatory simulation tools, participated in this project and collaborated in analyzing the agricultural economics (Ishida et al. 2007). For example, the last three decades have seen the expansion of upland cash crops using the rain-fed lowland rice ecosystem across upper northeast Thailand. The researchers thought that rice prices dropped while sugarcane prices remained high. This project aimed to validate such hypotheses, understand the decision-making processes of farmers, and identify how land should be used in the future.

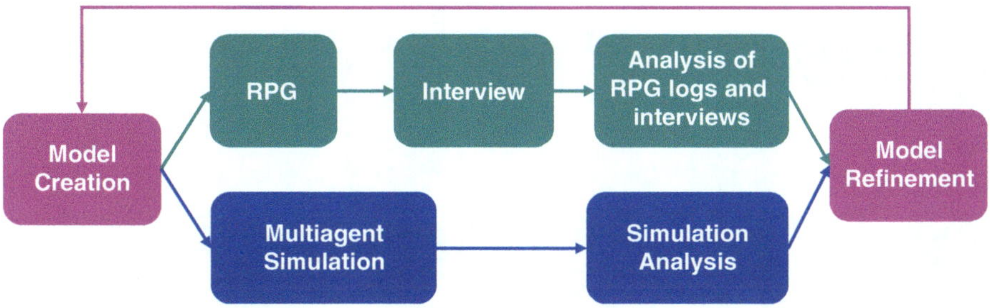

Fig. 6.5 Participatory process for economic analysis

Figure 6.5 illustrates the participatory process used at this time. An initial agent model is created from relevant literature and surveys. In contrast with the previously discussed evacuation and traffic simulations, when analyzing economics and politics, we use the method called *gaming* to model stakeholders. RPG (role playing game) sessions using a board that represents the stakeholders' environment were conducted. Next, the reasons behind the decision-making processes in the RPG were exposed by interviewing the stakeholders. Finally, the agent model was refined by analyzing the RPG log data and the interviews.

By running a number of RPGs, we improved the agent models, which were then used in multiagent simulations. The simulation interfaces have features similar to those of the RPG. The shared experience of modeling (RPG) enables stakeholders to accept the simulation results as what may occur in the future and propose model improvements. By using RPGs and multiagent simulations in this way, we achieved the participation of stakeholders and obtained an agent model that approaches reality.

Figure 6.6a shows the particulars of the RPG that reproduced the environment surrounding the stakeholder on the board. The results of the RPG were reflected in the agent model and then the land use simulation was run. Figure 6.6b shows a farmer, who had participated in the RPG, evaluating the validity of the agent model and the simulation; the RPG used the results of a simulation of planting crops both in the highlands and lowlands. The simulation exposed that (1) the farmers were aware of an incentive to grow upland cash crops, and (2) large-scale sugarcane quota leaders influenced the crop choice of smaller growers who did not have strong land-use strategies. The role playing games and the multiagent simulations raised the farmers' awareness of the effects of a decrease in sugarcane prices and the need for coping strategies. Here, too, the multiagent simulation was useful in creating a new social system.

It is worth noting that the agent model used here was not the state transition type used in the evacuation simulation. As shown in Fig. 6.7, the farmer's decision-making process was defined as a decision tree. Farmers select seed suppliers from the harvest of the previous year. They buy rice seed from Rice Research Center (RRC) or retail stores (ST), take out a loan with the Bank (BA) for Agriculture and Cooperatives or use their own stock seed (OS). Figure 6.7a is the initial decision

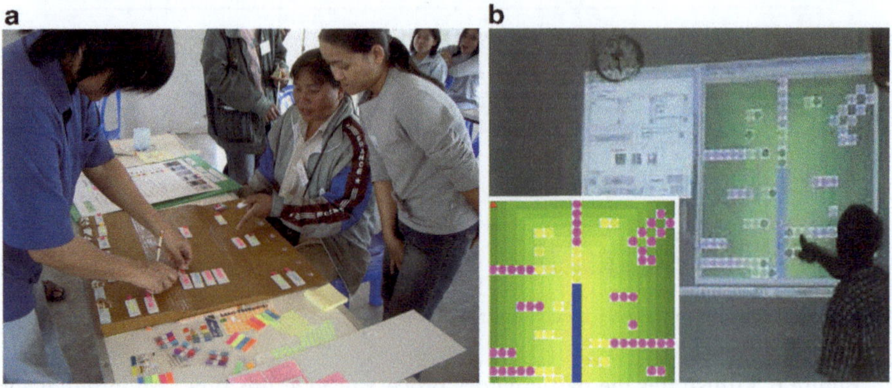

Fig. 6.6 Execution of simulation (**a**) role playing game, (**b**) evaluation of simulation result

tree model of the farmers. However, not all the decision-making processes were identified at this stage. Figure 6.7a was created as an initial model of the assumptions on farmers' decision-making processes from knowledge available at that point from specialists and the literature. Figure 6.7b shows the model created after a machine learning process where the initial model was refined using RPG data. The decision tree became more detailed so that unknown parts were clarified. Figure 6.7c is the final model improved through discussions of simulations between the domain experts and the farmers. In this way, the participatory approach improved the decision-making model and made it more reliable. The participatory process makes it possible to extract previously unknown factors that impacted the stakeholders' decision-making, and help stakeholders understand the real problems. Further, the RPGs made for a lively discussion among the participants themselves, making it possible for them to learn each other's ideas and plans for selecting the produce to grow. In this example, it was reported that the stakeholders accepted the simulation results as being realistic of future possibilities. Moreover, from this fact as well, we can see that participatory simulations are useful in providing the goal of stakeholder agreement. The participatory process demonstrates that due to the friendly nature of RPGs, even a community that is not accustomed to problem solving can adopt such measures.

6.5 Using Multiagent Simulations

Multiagent simulations can be used as predictive tools for spurring innovations in society and our daily lives. As examples, we discussed cases in which simulations were used in evacuation, traffic and economic analyses. However, when actually implementing a multiagent simulation, the points illustrated below need to be considered.

First of all, we have to be careful about how we interpret the results obtained. Let us suppose, for example, that when we run a traffic simulation a lot of accidents

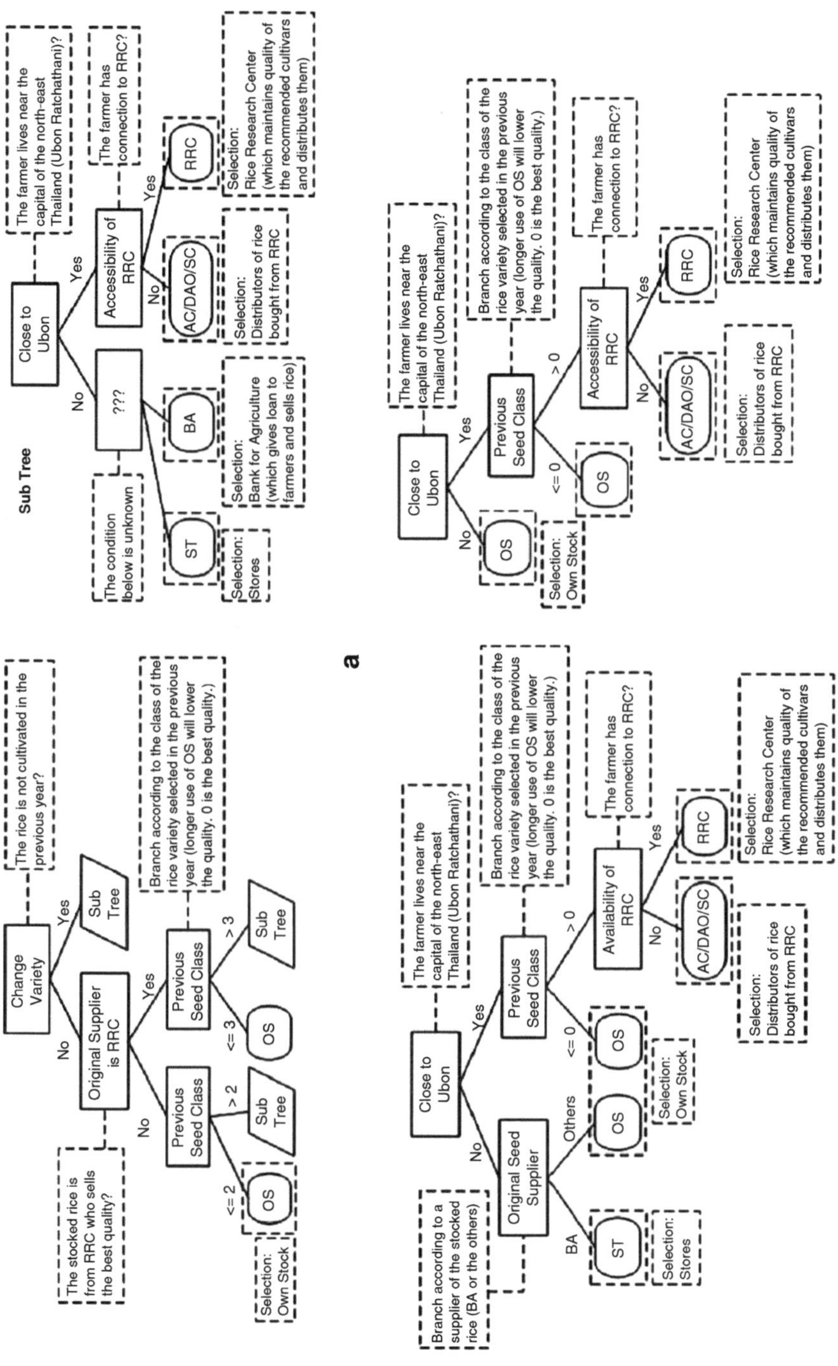

Fig. 6.7 Decision-making model of farmers (**a**) initial model, (**b**) RPG result, (**c**) final model

occur in the virtual city space. What does this result mean? Can we assume that people act identically in the virtual space as they do in the real world, or does the lack of embodiment introduce a fundamental difference with regard to our decision-making? Aren't we open to the criticism that we can get any result, depending on how we create our agents? In order to answer these questions, we have to be perspicacious in what is important and what is not in the results we obtain from our simulations. Time and time again we come across reports that handle the data obtained from simulations with all too little consideration, as if they were observation data from real life. The results of simulations are the results of agent modeling and environment setting, so in order to interpret those results, we have to examine the validity of the modeling and of the environment setting. In particular, we have to examine how the things we eliminate when modeling a phenomena impact the simulation results. Also, we should be able to explain the simulation results logically with a firm theoretical basis. In contrast to the natural sciences, when we get a result we cannot explain, it does not mean it is a new discovery. Before we conclude that a result is unexplainable, we have to analyze why it happened according to a theory.

Second, the purpose of running simulations is not simply limited to that of reproducing phenomena. For example, the agricultural economic simulation from Sect. 6.4.3 involved both understanding the current situation, while simultaneously looking to reach agreement on the design of future agricultural economics. As long as the ultimate aim of a simulation is to create a social system in the near future, then what is demanded of a simulation is not just the accuracy or generality of the results, but a means of satisfying needs we are not initially aware of. When stakeholders take part in a simulation, or when topics for discussion emerge from simulations, we can see needs we had not noticed before, and that can bring about new technologies and with them, reforms to society and how we live.

To readers new to the study of multiagent simulations, actually using a simulator must seem a shortcut to understanding. We are making the scenario description language that was used in the evacuation simulation in Sect. 6.4.1 available, linked with a network game, gumonji/Q. By modeling a problem you are familiar with and running a simulation, we think you will come to understand the functions and roles of the multiagent simulations discussed in this chapter.

References

Gilbert, N., Troitzsch, K.G. *Simulation for the Social Scientist.* Open University Press, 1999

Axelrod, R. *The Complexity of Cooperation: Agent-Based Models of Competition and Collaboration.* Princeton University Press, Princeton, New Jersey, 1997

Tadokoro, S., Kitano, H. *RoboCup Rescue — Challenges to Large-Scale Disaster Relief.* Kyoritsu Shuppan, Tokyo, Japan, 2000

Izumi, K. *Artificial Markets — Complexity Approach to Market Analysis.* Morikita Publishing, Tokyo, Japan, 2003

Nakanishi, H., Ishida, T. *Projecting Interaction in a Digital City.* Architectural Institute of Japan Journal of Architecture and Building Science 122(1565) 10–13, 2007

Hattori, H., Nakajima, Y., Ishida, T. *Learning from Humans: Agent Modeling with Individual Human Behaviors*. IEEE Trans. on Systems, Man, and Cybernetics — Part A: Systems and Humans 41(1) 1–9, 2011

Ishida, T., Torii, D., Murakami, Y., Terano, T. *Social Simulation and Participatory Design*. Information Processing 48(3) 271–277, 2007

Raney, B., Cetin, N., Vollmy, A., Vrtic, M., Axhausen, K., Nagel, K. *An Agent-Based Microsimulation Model of Swiss Travel: First Results*. Networks and Spatial Economics 3(1) 23–41, 2003

Chapter 7
Inclusive Design

Takayuki Shiose

Abstract User-participatory design (also known as participatory design), in which the opinions expressed by users actively influence the development of products or systems, has received a lot of attention in recent years. However, it is quite often the case that users are simply interviewed about their needs before the actual design process has even begun, or are only invited in to evaluate the usability of products which have already been developed and thus cannot be easily modified. Inclusive Design is a new design method which aims to imbue products or systems with more universal value by involving specific users as "lead users" throughout the entire product-development process, first by thoroughly examining those users' individual needs and then by developing, based on these needs, "multiple scenarios" which can involve a variety of other users (Wendy 2008; Clarkson et al. 2003). To give a few examples of Inclusive Design in use, such products as a new package design for drinking water and an easy-to-use nail gun for do-it-yourselfers have been developed by observing the actual situations in which people suffering from hemiplegia or rheumatism use the products (Clarkson et al. 2003). One of the remarkable features of this new approach to design is the proactive and deliberate inclusion of a wide variety of people as lead users, such as elderly and disabled users who had until now largely been excluded from the design process. Of course, Inclusive Design is not merely a means of drawing out the needs of minorities. What matters most is the simple fact that users do not have on hand a clearly defined, specific set of needs in the first place, much less the kind of vocabulary needed to determine or describe the value of, say, a new information service or piece of software that they've never before seen or touched. True innovation, that which is deeply involved in our society and in our lives, is the process through which users and system designers, working in collaboration, give shape to those needs which had not previously existed.

T. Shiose (✉)
The Kyoto University Museum, Yoshida-Honmachi, Sakyo-ku, Kyoto, Japan
e-mail: shiose@inet.museum.kyoto-u.ac.jp

T. Ishida (ed.), *Field Informatics*, DOI 10.1007/978-3-642-29006-0_7,
© Springer-Verlag Berlin Heidelberg 2012

7.1 User Participation in the Design Process

A well-known design model, called Universal Design, aims to create products which can be used by anyone, regardless of age, stature or the presence or absence of disabilities (Wendy 2008; Merton 1987; Preiser and Smith 2010). Examples of this design philosophy are already beginning to appear in such places as vending machines designed with coin slots installed at a height which makes it easy for people of any stature to buy the products on offer, or in the simple layouts of websites which anticipate the use of software for reading out onscreen text to visually impaired users. However, while it's all very well and good that products now exist which are easy for anyone to use, a commonly cited problem with this approach is how the overuse of the slogan "for anyone to use" often ends up confusing the people involved in product development. This is the reason why Inclusive Design, in which the design process begins by looking at the lifestyle of a single individual—the "life of John or Jane Doe," if you will—has been receiving so much attention. The Design Age Network, which was founded in Europe in 1991 to research the efficacy of designs meant to accommodate aging populations, formed the nucleus of the Inclusive Design movement, which would go on to establish a world-class research center at the UK Royal College of Art in 1999 (Preiser and Smith 2010). In Inclusive Design, new ideas are first worked out through the process of thorough and repeated interaction with and observation of individual users. Those ideas are then developed into multiple scenarios which anticipate situations involving a wide variety of different users. For this purpose, it is very important to engage in thorough dialogs with the individual users who participate as "lead users." The fact that we as designers tend to overly rely on knowledge and previous information, while neglecting to lend an ear to the real voices of users, becomes incredibly clear to us in Inclusive Design scenarios.

While we talk of making user-friendly products, the fact is that, in the world of engineering where things have to be made according to concrete terms, we simply cannot rely on subjective terms like "easy-to-use." While many people sympathize with the idea behind Universal Design, without a way to render concrete design procedures from it, it remains nothing more than an idealistic dream. In the 1980s, products began to be designed with a greater emphasis on ergonomics, that is, the study of the physical and physiological structure of human beings, including keyboards which minimize muscle strain and chairs which prevent lower-back pain. In the 1990s, with the increasing need for ease-of-understanding in computerized products, development began to focus mainly on designing products based on cognitive engineering, which focuses on the cognition and behavior of human beings. At present, the proactive introduction of a systematized method for both experts and users to evaluate products, called usability evaluation, has ushered in an age in which products can be evaluated according to more subjective terms, such as usability or satisfaction. Research is currently being conducted into new methods of product development, such as the Persona Method, which create a

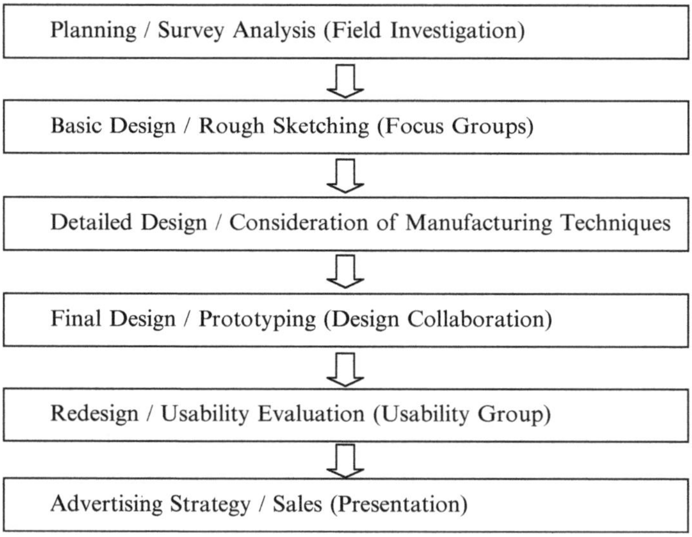

Fig. 7.1 Stages of the design process in which users participate

virtual user, detail his or her virtual lifestyle, and then describe that user using products in various specific situations (Wendy 2008).

However, if we are too hasty in trying to translate this method into the *lingua franca* of the engineering world—efficiency—we may end up with nothing more than virtual users designed on the assumptions of experts, which can be a far cry from real-world user models.

In order to obtain a broader view of not only usability but also the meaning that products have in the lives of users, we must go beyond basic design, which aims to determine what consumers want, to the point where we can perceive what an effective, efficient, satisfactory product is in the first place.

Figure 7.1 shows how, in a typical product design process adopting User-Participatory Design, opportunities for user participation are limited to usability groups and the product refinement stage. However, in Inclusive Design, users are also given opportunities to participate in the early stages of the design process, such as field investigations and focus groups. The establishment of this method, allowing users to take part in the entire design process from beginning to end, is the reason why Inclusive Design has been getting so much attention.

However, due to the high level of expertise required, there have thus far been few successful examples of user participation in the stages of the design process which work out more minute details and consider manufacturing techniques. Still, new techniques which make manufacturing technology more accessible to consumers and non-specialists, such as Desktop Factory, are also currently under development (Clarkson et al. 2003).

In the realm of product and system development, there is now a pressing need to train people so that they are capable of understanding the Universal Design method and making proper use of concrete usability evaluation techniques. Furthermore, it is important to diversify the ways in which users participate and to perfect the methodology of Inclusive Design, in which users' opinions are directly linked to new and innovative designs. Consequently, such measures will open up opportunities to collaborate with users to address currently unseen needs related to, for example, information services or software.

7.2 Awareness in Design Workshops

7.2.1 The Inclusive Design Workshop Process

The main feature of Inclusive Design is that lead users can actively participate in the early stages of the design process, such as survey analysis and basic design (Clarkson et al. 2003; Shiose et al. 2007). Figure 7.2 gives a simple outline of a typical, 1-day Inclusive Design workshop, divided into four stages. Each design team is made up of a group of six to eight participants, and there are four or five such groups, making a total of 30–40 participants in a workshop. People with various disabilities, such as those with visual or hearing impairments, hemiplegia suffers, wheelchair users, and the elderly, are invited to act as lead users in each group. The remaining participants in each team include designers, engineers, researchers and students covering various fields of expertise. The participants include experts from a wide range of fields, with designers who are active in everything from product design to textile design and engineers who come from industries ranging from the basic manufacturing of automobiles or home appliances to information technology. The overall creativity of a workshop is directly related to the diversity there, rooted in the expertise of the participants.

1. The fieldwork stage (Fig. 7.2a) is not limited to conducting face-to-face interviews; everyday-life activities, such as buying tickets at a train station or shopping for daily necessities, are enacted by the team members in the real world. Each member is asked to take thorough notes on the things he or she notices. At this point in the process, every little observation or relevant notion is simply put into words one after another, without regard for category or importance.
2. In the idea-sketching stage (Fig. 7.2b), the items of awareness obtained during fieldwork are collected, laid out and arranged on paper or whiteboard. While quickly sketching out these observations, team members also develop them into more concrete sketches of ideas while taking into consideration multiple scenarios which can involve various users. One of the main features of this approach is the generalization of user models based on the degree to which needs are held in common.
3. In the prototype-production stage (Fig. 7.2c), idea sketches are reproduced in as close to life-size as possible. Because it is more important for the members to

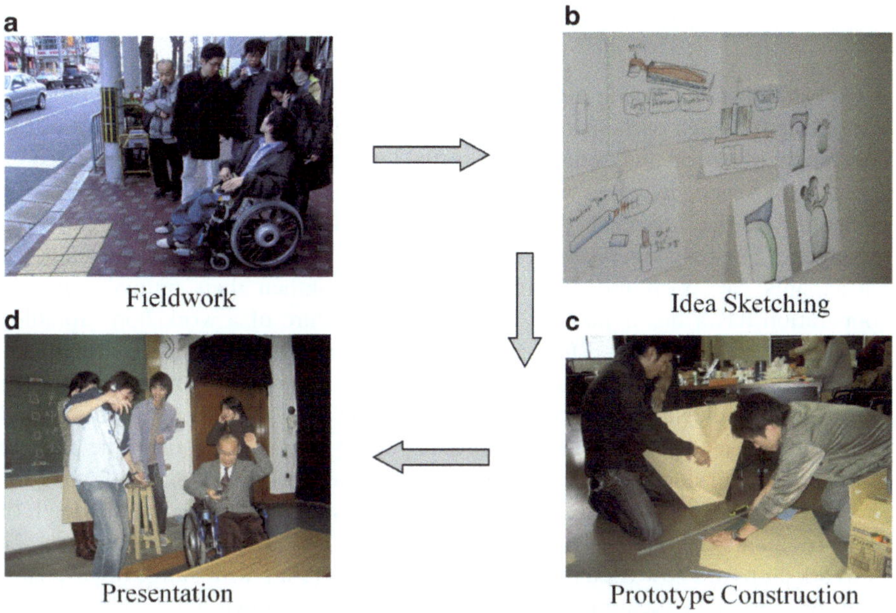

Fig. 7.2 Stream of events in an inclusive design workshop

discuss the functionality and structure of a prototype than it is for them to pursue accuracy in specific details, the key term here is "Quick & Dirty." At this stage in the process, by making an idea into something with substance, physical and structural limitations make the parts of the idea which had been ambiguous more concrete.

4. In the presentation stage (Fig. 7.2d), each team is asked to make a presentation, which focuses attention on the relationship between the awareness raised and the ideas subsequently developed, to the members of the other teams. The very act of rearranging one's own ideas, in the form of words used in a presentation, can deepen one's understanding of user needs. This stage is the act of logically arranging words into sentences and paragraphs which describe the entire process from beginning to end: from finding out the concrete needs of an individual user through dialog, to creating a generalized user model from the overlap of users' needs, to coming up with an idea and making it concrete by observing its physical and structural limitations.

There is now a greater need for a participatory design format in which lead users can make an even greater contribution to the product design process as full-fledged members of design groups. Lead users should participate in focus groups from the basic design stage of development and propose specific design ideas alongside developers and engineers. This kind of participatory design format is distinctive in that it involves users in the process at the same level as designers, making it possible to develop expressions of user needs into something both more concrete and more positive.

7.2.2 Why Workshop?

A workshop is not a setting for its host to pass on information unilaterally; rather, it is a setting in which participants take center stage, participating actively not just to understand things on an intellectual or verbal level but to also learn things based on actual experience. In cases where a group of people who have never met are gathered together, a workshop will usually start off with a small game, called an ice-breaker, to help everyone feel more at ease. Whether or not the gathered participants can open up to each other and strengthen their mutual sense of cooperation has a big impact on the success or failure of a workshop (Spradley 1980). The person assigned to lead the workshop and thus guide the participants closer to discussions of core issues, is called the facilitator. It is necessary for the facilitator to dedicate himself or herself to relieving tension among participants and to creating a friendly atmosphere while also paying careful attention to the passage of time to ensure steady progress in keeping with the workshop's planned schedule.

The facilitator's skill in running the workshop is directly related to the depth of the participants' learning.

Workshops are now used in various situations and for a wide range of purposes, such as formal education, business meetings and town-hall style citizen-participatory community development. It is especially important in workshops to fight the tendency toward knowledge-ism—placing too much value on the opinions of experts rather than the opinions of ordinary people—and to get all of the participants to understand the importance of working together in a shared space to put into words what they feel and want. It is not enough for elderly or disabled people simply to express their opinions; rather, what matters most is the putting into words of things which people had not previously been able to express on their own. In a design workshop setting, this is accomplished through such universal activities as working together to create something that is concrete. One good example is a workshop in which nurses participated with the aim of designing better hospital slippers. In the course of the workshop, the participants realized just how ambiguous one simple phrase they'd been using, "take care to prevent patients from 'falling down' in the hospital," could be. The nurses had thought that they all shared the same meaning when they said the words "fall down." However, when attempting to apply their ideas about nursing care to a concrete design for hospital slippers, it became clear to them that they had actually only shared an ambiguous definition of "falling down," with meanings varying from "stumbling and falling due to excess friction," to "slipping and falling due to insufficient friction," or even "losing balance and falling due to slippers coming off." This is an example of just how ambiguous our human languages can be, and it also illustrates the necessity of discussing such minutiae in order to understand needs in concrete terms.

7.2.3 The Systematization of Workshop Management

The curriculum of a workshop depends to a great extent on the ability of the facilitator, who, for better or worse, runs the entire show. In order to provide participants with reliable learning experiences continually, and to encourage the most widespread use of workshop activities, it is necessary to run workshops systematically and, at the same time, to establish a method for accumulating workshop outcome case reports. It is also necessary to select the workshop format which is the most effective for learning from among various types, considering everything from the number of participants, the combination of participants' specialties, the amount of time available, and the theme of the workshop itself (Wendy 2008) (Clarkson et al. 2003; Shiose et al. 2007).

Table 7.1 illustrates seven real-world examples of workshop formats with varying lengths of time, design targets and user participation styles (WS-I to VII).

The themes of the workshops are set in advance in WS-I, II, V, VI and VII; in WS-I, V and VI, they are related to product design, while in WS-II and VII, they are related to the design of information systems. Facilitators also varied: some are product designers, while others are graduate students majoring in informatics.

In both WS-I and VI, the lead users include people with cerebral palsy, wheelchair users, and people with visual impairments, and the target product is the kind of electric water pot used in everyday life, usually to heat water for tea and other beverages. For example, lead users in wheelchairs describe the difficulty they experience in using electric pots because the pour button is located too high for them and the movable handle is too unstable to carry the pot into the kitchen to fill with water. Figure 7.3 illustrates an instance of an awareness made in WS-VI and an example of a design idea created from that need. For visually impaired lead users, in order to plug a disconnected power cord into the pot's connector, they have to feel around the base of the pot with their hands. At this point, the design team came up with the idea of fitting slit-like vertical indentations around the base of the pot. By making the slits come closer together the nearer to the connector one felt, it became possible for users to locate the connector easily by simply rubbing the base of the pot. In addition, from the awareness of another visually impaired lead user that the pour button was difficult to press, regardless of whether one has a disability, the need emerged for an easier-to-press button. Also, from the awareness of a lead user with spina bifida, the need emerged for the lid's opening and closing mechanism to require less upper-arm motion. Thus, by adding a big, easy-to-press button resembling a face and a clip in the shape of an ear to open and close the lid, the idea for the Pet Pot came to fruition.

WS-III and IV are for longer-term workshops, and they require substantial time for lecturing and guidance on Inclusive Design. The design target is not set in advance but determined through communication with people having disabilities over the course of the three-day workshop. For example, one of the groups in WS-III observed that disabled people had trouble with video arcade games in which players compete using agility, such as fighting games or racing games.

Table 7.1 Examples of inclusive design workshop formats

Format	WS-I	WS-II	WS-III/WS-IV
Duration	Half-day, 120 min	Half-day, 150 min	3 days, 21 h
Site area	25 m^2	25 m^2	60 m^2
Design theme	Electric pots	Infrared radar	Enjoyable living
Lead users	Users with cerebral palsy/physically disabled users	Visually impaired users/users with cerebral palsy	Visually impaired users/hearing-impaired users/wheelchair users
Facilitator	Designer	Graduate student in field informatics	Product designer
Format	**WS-V**	**WS-VI**	**WS-VII**
Duration	Half-day, 180 min	2 days, 12 h	3 days, 20 h
Site area	15 m^2	30 m^2	30 m^2
Design theme	Picture books	Electric pots	Information services
Lead users	Hearing-impaired users	Visually impaired users/users with spina bifida/wheelchair users	Visually impaired users/wheelchair users
Facilitator	Graduate student in mechanical engineering	Product designer	Graduate student in field informatics

Fig. 7.3 Awareness obtained about a visually impaired lead user's electric pot usage

This awareness led that group to focus on more communication-oriented games, such as *purikura* photo booths in which people take pictures with friends. From there, the group took a step further and, anticipating the needs of visually impaired players, suggested an idea for a rhythm harmony game which anyone could play. Another group suggested various ideas for coffee drippers which would make it easier to make drip coffee for visually impaired users or those with the use of only one hand; these ideas included, for example, designing one section of the coffee filter to be thinner. Each of these designs, which came about as the result of considering multiple scenarios, endows its target product with benefits which can be appreciated regardless of whether the user has a disability.

In WS-II and VII, visually impaired people and wheelchair users are invited to participate as lead users, with the design theme being new devices using infrared sensors to detect one's distance from the nearest wall and other such navigation systems. One observation, for example, was that if a visually impaired person were to get lost around buildings, he or she would have no way of knowing his or her location or of relaying nearby landmarks to somebody over the phone. From this awareness, ideas were suggested such as a portable device making use of GPS to allow one to share audio and visual information about nearby landmarks over the phone. However, participants found it difficult to share ideas about novel new devices that they had never before touched and information services that they could not see. Ultimately, this workshop established as an issue for future consideration the matter of how to bridge the gap between awareness and suggestions relating to systems.

7.2.4 Design Workshop Management

The outstanding characteristic of an Inclusive Design workshop is the involvement, to the extent possible, of lead users in the entire design process, from beginning to end. When it comes to designing an electric pot, working together with visually impaired people and wheelchair users in actual product usage situations allows designers to understand how the difficulty of using such products must be skillfully circumvented through arduous and painstaking motions in these users' everyday lives. When it comes to designing an information system, hearing-impaired users can help designers to realize that an information display terminal which is hard to follow with the eyes is difficult for people to read, regardless of whether the user can hear. And by having actual designers and engineers present, it is safe to expect that suggestions for improvement will actually be applied in future product design.

At the same time, the process of designing products by collaborating with visually impaired people or wheelchair users dispels preconceived notions on both sides and alters the mutual relationship between designers and lead users from being about designing products "for" disabled or elderly users to being about designing products "together with" disabled or elderly users. The flow of this kind of workshop management process is illustrated in Fig. 7.4.

(1) The design theme in a workshop can be set to anything from concrete products, such as electric pots or umbrellas, to more abstract concepts, such as lifelong learning or playing. The easier the theme is to associate with something concrete, the easier it is for people to tackle the design process and to enjoy a sense of accomplishment. However, at the same time, because more abstract needs are difficult to work into design targets, there is a need to increase trial workshops which incorporate these themes in order to develop more innovative design systems.
(2) Next, one must work to include as wide a variety of users as possible. In a workshop where the theme is umbrellas, normal umbrella usage will entail something entirely different for visually impaired people than it does for wheelchair users, and their demands will of course be very different as well. In a workshop where the theme is "bicycle policy," the demands of elderly cyclists will be completely different from those of mothers riding with infants.

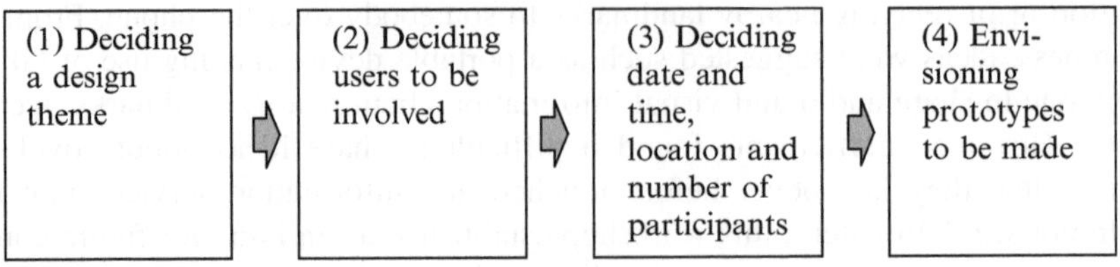

Fig. 7.4 Management flow of a design workshop

Increasing the number of lead users gives us more opportunities to incorporate new points of view into the design process.

(3) When setting the date and time for a workshop, it is important to keep in mind that convenient times can vary from participant to participant and from weekday to weekend. When selecting the location for a workshop, sites where participants can experience the day's design theme to the maximum extent are desirable. For example, if the design theme is electric pots, a site with a fully outfitted kitchen is preferable; if the design theme is museum displays, a site inside an actual museum is preferable. It goes without saying that one must secure an adequately wide space for 30 or 40 people (including lead users) to be able to work while engaging in an open-minded exchange of opinions. When wheelchair users are involved in the workshop, the bare-minimum essentials include confirming the availability of handicap-accessible toilets and elevators and verifying that the site has an entrance ramp.

(4) Finally, one needs to have an advance mental image of what kind of prototypes are to be made at the end of the workshop in order to select the appropriate materials for making mockups. If the theme is some kind of large home appliance, one may use Styrofoam boards or cardboard to create models in as close an approximation of life size as possible. If the theme is website design, one can consider such options as whether to use PowerPoint or other PC-based presentation software or mockup software, or whether to use paper prototyping, in which participations create picture card style page views out of letter-sized paper. In addition to aiding in the dialog among lead users, designers and engineers, the act of making something take form before one's own eyes is extremely important. Therefore, the selection of mockup materials can influence whether or not the workshop is a success.

7.3 Design Workshop Preparation and Step-by-Step Process

7.3.1 A Product Design Workshop Case Study

First, as an example of a workshop in which the design theme is a concrete product, let us take up the umbrella. If there existed one which was "large yet light when in use and able to be quickly folded into a very compact form when not in use," who

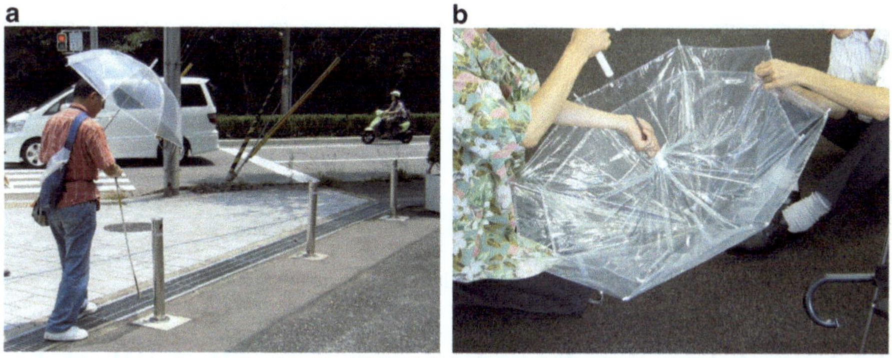

Fig. 7.5 A design workshop with the theme of umbrellas. (**a**) A scene from fieldwork. (**b**) A dual-structure, sound-suppressing umbrella prototype

wouldn't want such a wonderful umbrella? However, although these kinds of abstract needs may very well apply to a wide array of users, they cannot lead to concrete designs, and any ideas conceived will invariably end up being nothing more than daydreams. Through Inclusive Design workshops, we can look for solutions based on more concrete needs.

Imagine, for instance, a visually impaired person or a wheelchair user participating as a lead user. For a visually impaired user, there are several problems involved in using an umbrella. First, he or she must hold the umbrella with his or her non-dominant hand because the dominant hand will be used to hold a walking stick. Second, the sound of raindrops striking an umbrella makes it difficult to listen to the environmental sounds of the surroundings, upon which such a person usually relies when walking on roads. Wheelchair users also have several problems. Many of them prefer to use a raincoat instead of an umbrella because they cannot hold an umbrella and move at the same time, and umbrellas typically aren't big enough to cover an entire wheelchair. What's more, because of the difficulty and annoyance involved in putting on a raincoat, many wheelchair users make every effort to avoid going out on rainy days altogether. The need for an umbrella which can be used without using one's hands is not limited to these lead users—it can also be a need for, say, people who want to bring their young children along when they go out.

Figure 7.5 illustrates scenes from an actual workshop held on the theme of umbrellas. A fieldwork exercise was first conducted by having a user with a walking stick recreate the actual situation of walking outdoors while carrying an umbrella (Fig. 7.5a). A dual-structure vinyl umbrella was designed to act as a sound suppressor for preventing environmental sounds from being drowned out by the sound of rainfall, and then mockups embodying this concept were created (Fig. 7.5b). In order to allow concrete ideas for umbrellas to bear fruit, it is very important here to prepare several inexpensive vinyl umbrellas in advance and instruct each group that they may work with them however they see fit. In workshop situations, the act of simply using one's own hands to give shape to the ideas which spring forth, without stopping to reject them as invalid, helps to solidify solutions to previously vague needs.

Fig. 7.6 A design workshop with the theme of information navigation

7.3.2 An Information-Based Navigation System Design Workshop Case Study

Now, let us imagine a workshop in which the theme is an abstract information service, namely, "guidance." It's much easier for participants to make prototypes and for lead users to recall individual experiences when dealing with the design of something concrete. However, with every imaginable industry in this day and age making the switch to digitalization and information-orientation, most of the information services we're used to have been replaced with systems which can't be seen or felt. Even when we try to incorporate the opinions of users in the design process, for a user who only uses a PC for simple word processing, or for a user who has never even touched a computer, the stream of technical terminology being spoken by IT system developers can sound like a foreign language. The creation of a new common vocabulary, which users, designers and engineers alike can use to express shared meanings, is one of the important, over-arching themes of Inclusive Design workshops.

Figure 7.6 illustrates scenes from an actual workshop on the theme of information navigation. A test was contrived for an information service (Fig. 7.6a, b) that addresses the situation of a visually impaired user getting lost and needing to receive directions from a friend over a cellular phone. Here, an issue came to light—namely, that it is difficult for a visually impaired user to actually ascertain the landscape of his or her surroundings in order to figure out the current location. Other valuable insights based on awareness were gained during this test. For example, one awareness was that it is extremely important for a wheelchair user, when approaching a building, to know in advance whether that building has an access ramp. Another awareness was that, due to a wheelchair user's lower line of sight, landscape features which another, taller, user could easily point out might be hidden from the wheelchair user's view behind buildings or other obstacles. Consequently, a mockup was created (Fig. 7.6c) of a prototype computer-based service which gives verbal route information via an automated response system. One awareness gained here was that, when one kindly attempts to provide a large amount of detailed information, the details lead to an explosion of information the second they are converted into writing. This kind of problem, which is related to our cognitive resources, is a hurdle which anyone may face, regardless of having a vision or hearing impairment.

7.4 Design Workshop's Viewpoints on Analysis and Information Assurance

7.4.1 Viewpoints on Analysis

In approaching potential, latent needs, methods such as ethnography are useful. Here, we introduce two examples of interview methods which are considered to be especially useful in Inclusive Design workshops (Shiose et al. 2007; Toffler 1984; Chambers 2002).

- Focus Interview
  This method uses a succession of questions which gradually narrow the range of possible answers by shifting from unstructured questions, such as "Was there anything in this system which left an impression on you?" to more specific, semi-structured questions and, finally, to highly structured questions such as "Were you able to clearly associate the information you were provided with the next operation you should perform?"
- Unstructured Interview
  In this method, after giving an advance explanation about the interview and its purpose, one extracts answers through natural conversations with people one meets in the field. This is a method for determining, through descriptive and comparative questions asked during a friendly, candid chat, how an informant (information provider) views the world, what makes him or her happy, what kinds of actions he or she takes and, forming the background for all of this, his or her culture and the lifestyle he or she lives.

7.4.2 Viewpoints on Effective Information Sharing

To manage a design workshop in which a wide variety of users participate, it is important to carefully consider the facilities to be used. For example, one must take into consideration participants' safety in traveling to and from the site, the ease of understanding where everyone is to assemble, the size of the site, and the availability of handicap-accessible toilets. Another matter requiring the utmost care and attention is related to information assurance: since a wide variety of people with various impairments, ranging from vision to hearing to physical impairments, are participating, it is necessary to maintain a dialog with the participants involved in workshop fieldwork and to constantly make sure that everyone understands what is happening during design work. By making each participant in a design workshop personally aware of the need for information assurance, it is possible for them to establish relationships on even ground with users, whom they can relate to as partners. However, when findings from awareness have been collected and participants begin to sublimate them into ideas, it is easy for them to become absorbed in the process of coming up with ideas and, without being aware of it, become less sensitive to information assurance for users. Statements filled with demonstratives, such as "How about making it bigger here" or "Let's put this and this together," only confuse visually impaired users. Hearing-impaired users can be left feeling alienated by the laughter of participants omitted from summary dictation notes. This is the kind of discrepancy in communication which can occur in situations where any user, regardless of the presence or absence of disability, engages in information sharing with designers and engineers.

While information assurance is very important for visually or hearing-impaired people, its methods of providing information also raise some thought-provoking questions about effective techniques for sharing information. For example, in the case of computer dictation—in which a typist enters the words of a speaker into a computer as letters—a skilled typist who is able to enter about 50–60 words per minute is able to transcribe about 70 % of what a slow-speaking speaker says. However, opinions are mixed when it came to "thorough information assurance," that is, transcribing into letters, to the extent possible, every sound which enters a typist's ear, from idle chatter to statements relating to design ideas. Some users had positive things to say, such as, "I felt like I could fully participate in group activities from which I usually feel left out." On the other hand, other people had more negative comments, such as, "It was like having movie subtitles appear right in front of my eyes. I felt like I understood what was being said, but my eyes got strained. Sign language interpreters are easier to understand because they know how to summarize properly." This very thought-provoking example makes us realize that, for hearing-impaired people, simply transcribing everything into letters isn't necessarily sufficient for information assurance.

Discrepancies involved in communication, for example between experts and non-experts, will be discussed in detail in the chapter on Outreach Communication. For effective information assurance, it is necessary to consider in greater detail what one should try to convey and what is, in fact, conveyed to a recipient, as well as what the difference between the two implies for both sides.

7.5 From "Design for All" towards "Design with All"

In this chapter, we introduced Inclusive Design as a technique for designing new products and systems. In this technique, we begin by taking a deep look into the lifestyle of an individual user, expand from there to include various other users in multiple scenarios, and then apply what is learned to the design process. The process of making people's ambiguous needs regarding merely imagined systems or services into something more concrete, and of doing so cooperatively through dialog among users, designers and engineers, is intimately related to the very process of creating innovation in our society and in our lifestyles.

Although the main characteristic of Inclusive Design is the proactive inclusion of elderly or disabled people as lead users, it is by no means simply design for the sake of minority users. Inclusive Design is receiving attention as a method for approaching more essential needs, those which involve a wider range of users. The term information assurance has a certain ring to it which tends to be perceived as a service to users with sensory impairments. However, it is in fact a necessary element for actively involving not only the lead users but all of the other members participating in the design work. The endless stream of technical terminology coming from an expert in a certain field can end up being an impediment to most of the other participants, regardless of whether they can see or hear. In envisioning a situation in which a designer will welcome, as partners, as diverse a group of users as possible, information assurance is one of the most important elements to consider for understanding what each design partner means by the words he or she speaks and how those words are perceived. Information assurance is also necessary for applying, to the extent possible, the experience and knowledge of all participants in a design workshop.

References

Wendy, A.J. *Universal Design for the Home: Great Looking, Great Living Design for All Ages, Abilities, And Circumstances.* Quarry Books, Minneapolis, Minnesota, 2008

Clarkson, J., Coleman, R., Keates, S., Lebbon, C. *Inclusive Design — Design for the Whole population*, Springer, 2003

Merton, R.K. *The Focused Interview and Focus Groups: Continuities and Discontinuities*, Public Opinion Quarterly, 51, 550–556, 1987

Preiser, W., Smith, K.H. *Universal Design Handbook, 2nd ed.* McGraw-Hill Professional, 2010

Shiose, T., Toda, K., Kawakami, H., Katai, O. *Inclusive Design Workshop by Regional Cooperation Between an NPO and a University.* International Workshop on Intercultural Collaboration, Springer, LNCS 4568, pp. 355–367

Spradley, J.P. *Participant Observation.* New York, Holt, Rinehart & Winston, 1980

Toffler, A. The Third Wave: *The Classic Study of Tomorrow.* Bantam, Reissue, 1984

Chambers, R. Participatory Workshops: *A Sourcebook of 21 Sets of Ideas and Activities.* Earthscan Publications Ltd., illustrated edition, London, 2002

Chapter 8
Participatory Design

Yutaka Yamauchi

Abstract Over the last three decades, participatory design research has reached maturity with a number of studies reported and a variety of methods developed and tested. This chapter has two goals. First, it is aimed at showing that there is more to be done in participatory design research. Stakeholders not only participate in design but also transform themselves through design. They start to see the world differently and design a technology in new ways. This complex participation process has been glossed over. Despite the plethora of research that has been done, we have yet to comprehend stakeholder participation fully. Second, it seeks to point out that participatory design is needed for a broader range of technology development, particularly upstream research. Whereas participatory design has been discussed in the context of downstream development of information systems and consumer products, it has become important for upstream research, which tends to be removed from the field in which the technologies will be used. Through these arguments, more research is called for in participatory design.

8.1 What Is Participatory Design?

Participatory design is aimed at including users in the design process so that users' opinions and preferences are reflected in the design of a technology they use (Bodker et al. 2004; Greenbaum and Kyng 1991; Schuler and Namioka 1993). Several methods of participatory design are listed in Table 8.1. The core method of participatory design includes workshops and design sessions in which users are encouraged to think creatively and propose their own ideas. Ethnography is also used in conjunction with participatory design to gain a deeper understanding of users.

Y. Yamauchi (✉)
Graduate School of Management, Kyoto University, Yoshida-Honmachi, Sakyo-ku, Kyoto 606-8501, Japan
e-mail: yamauchi@gsm.kyoto-u.ac.jp

T. Ishida (ed.), *Field Informatics*, DOI 10.1007/978-3-642-29006-0_8,
© Springer-Verlag Berlin Heidelberg 2012

Table 8.1 Methods of participatory design

Method	Description
Workshops	Users and designers collaborate to create a design, vision, policy, or understanding of the current setting jointly in a focused manner
Ethnography	Designers conduct an in-depth observational study of users to obtain a first-hand understanding of users' circumstances
Cooperative prototyping	Users experience potential technologies with a prototype and modify the prototype cooperatively along with designers
Mock-ups	Designers create mock-ups, often using cardboard, to stimulate users to think about the potential technology and new work practices
Card sorting	Designers learn the users' information environment by having the users write down types of information related to cards and group them into piles
User design	Users are given a design tool to create a design on their own. The design tool must be sufficiently easy for non-programmers to use

Beyond workshops, ongoing activities in which designers prepare prototypes and mockups and obtain feedback from users in a continuous manner are also crucial. To assist users who are not familiar with design practices, cards are often used as a medium of gathering and structuring data. In a more proactive approach, users are given design tools with which they can explore designs on their own.

This chapter presents discussion of the basic rationale for stakeholders' participation in design beyond the ethical issue of involving users in decisions that affect them and the practical issue of tapping into users' extensive domain knowledge. Through this discussion, it is then argued that stakeholders' learning is critical but challenging. I highlight the issue of power inherent in this learning. Then, I seek to expand the scope of participatory design beyond involving stakeholders in design towards any kind of technological design that incorporates a social as well as technical system.

8.2 Why Participation?

Design is political (Langdon 1980). Any design is aimed at a desired end: what ought to be rather than what is (Simon 1968). Furthermore, it is somebody's desired end. Designers of a technology are in the position to dictate what users can and cannot do. It is true that in any social interaction, we have expectations of what others can and cannot do. Yet, technology design is particularly powerful in this capacity. Although technology can accomplish what would otherwise be impossible, it can also limit people's behavior. Stakeholder participation is therefore an ethical issue.

Technological design is also complex (Simon 1968). It requires extensive knowledge of the domain. Designers, however, typically have little knowledge of the particular domain. They must learn a lot. Ethnography is often used to immerse oneself in the users' environment and gain a deeper understanding of the users and other stakeholders. User participation is proposed as a means to tap into users' domain knowledge.

Furthermore, participatory design is called for because users must have a sense of ownership to make better use of the technology. Workers who can exercise certain control over their work tend to be more motivated and tend to develop more efficient and effective work practices. Technologies cannot produce value on their own: users' practices deliver value. Some users might simply use a limited set of functionalities and ignore the potential value of a technology, whereas others can adapt to using the technology creatively (Yamauchi and Swanson 2010). User participation is an effective means to empower users.

However, a deeper reason exists for user participation. Design of a technology encompasses not only material artifacts but also people. Lives of the users will be affected by the technology. Unintended consequences might affect the lives of many others who do not necessarily touch the technology. This situation necessitates that we conceive the technology within the socio-technical system, a system that encompasses technical and social conditions (Trist and Bamforth 1951).

Designing a socio-technical system differs from designing a material artifact in isolation. Material artifacts behave according to physical *laws* such as Newton's laws. Although natural laws are not explicit, information technologies are deterministic in that their behaviors are predicted by specified codes. Designing a social system is a completely different matter. Designers cannot dictate completely how humans behave, primarily because human behavior is governed not by *laws* but by *rules*. Rules differ from laws in that we can always violate them. Although natural laws are always falsifiable and are valid only under certain conditions as Thomas Kuhn (1962) showed, material artifacts do not violate the laws under those conditions.

Rules differ from laws because people who follow the rules have a certain understanding of them. A simple rule can be considered: Stop at a red light. When following this rule, we know that we are expected to follow it. In this sense, most people have a reflexive understanding of the rule. Furthermore, everyone is held accountable for following the rule. When we start moving at a green light, we expect others to stop at their red light. Rules are always normative in that we are punished when we violate them. Not only might we be stopped by a traffic officer, we are also seen by others to have violated the rule and are thereafter regarded as reckless.

Rules are reflexive. Therefore, we cannot simply create a rule and expect people to follow it blindly. For instance, in the early days of American football, injuries became a concern. Then helmets were introduced to protect players from injuries, with an expectation that injuries would decrease. Astonishingly, injuries increased thereafter. Because the players had additional protection, they started to play more aggressively and more dangerously. This simple example illustrates the difficulty in designing human behavior. For professional players, it was important to show flashy displays of skill and daring. Without understanding this, we often fail to achieve an expected result. To design a whole socio-technical system, it is crucial to know the reflexive understanding that people have rather than simply imposing constraints and procedures.

Involving only stakeholders in the design discussion is insufficient. If we had asked football players how to reduce injuries, they would probably not have given any viable design. Stakeholders themselves tend to have overly simplistic understanding. Most stakeholders' reflexive understanding of the socio-technical system is not a conscious thought that they can readily express (Giddens 1984). Moreover, we cannot design something that will work in the first attempt. Unanticipated consequences invariably arise. Design entails constant trial and error, or reflective conversation in which we try something and to determine whether it works (Schön 1983), which involves iterative design through which learning accumulates. When we try one idea, it might not work. Therefore, we learn and try a different approach. Design must be tested and validated constantly.

Beyond simply involving stakeholders in design decision making, design must be part of every stakeholder's life. Stakeholders must learn how to design the socio-technical system in which they find themselves. It is, however, not easy for stakeholders to learn to design, as discussed in the next section. Furthermore, to accept this notion of participation—beyond simply involving stakeholders in design, one must expand the scope of participatory design. Typically, participatory design has been applied only to situations with a definite group of users. Any technical design, which almost always involves social design, must be participatory. Therefore, participatory design is discussed below in a research setting, where researchers must consider and deal with stakeholders' reflexive actions.

8.3 Stakeholder's Learning

8.3.1 Learning and Transformation

Users and other stakeholders must learn and transform their own knowledge to start contributing to design. When asked what kind of a technology they want, users typically suggest what they have been using and recommend possible improvements related to some annoyances that they have been frustrated with. Two problems are inherent in this situation. First, if these requirements are all realized, the development cost will almost invariably be much larger than the budget. Even given a very generous budget, the system will usually be too enormous in scale to maintain. Some requests are only "good to have" whereas others are "must have." It is difficult for users themselves to distinguish these two. They must learn to take a more holistic perspective. Second, such incremental changes do not justify costly technology development. Technology development is proposed to achieve a more significant improvement and change. Faced with this gap, stakeholders must transform their knowledge, break out of their familiar reality, and design a technology to achieve larger goals.

To make the situation even worse, stakeholders are typically busy with their own daily work and cannot easily find time to participate in design. In most cases, they do not want to participate because they think that technological development is somebody else's job. They typically say to designers, "I tell you what we want, and you can then make it." In a typical information system development, a client contracts with an IT vendor. Designers of the IT vendor will work with another client after they are done with their current one. In contrast, users will continue working as users and will not produce another information system design in so many years. Therefore, stakeholders do not think that design is their responsibility: They face the initial challenge of understanding that they must design the system they will use.

The difficulty in learning is that a certain view of the world is ingrained in stakeholders. They take the world they know for granted and take it as the reality (Berger and Luckmann 1966). In participatory design, then stakeholders must challenge their own worldview and obtain a new perspective. Nevertheless, this process is not straightforward. Realizing that what one has thought to be the reality is not so grounded results in despair. For designers who must help stakeholders design, it is insufficient to listen to them. It is often necessary to dispute what stakeholders request and then convince them that something different would be better.

In information system development, most system designers understand that stakeholders' participation is vital. The likelihood for the stakeholders to resist the system increases if they do not feel a sense of ownership. More than that, when testing the system with actual work situations, implementing the system in the workplace, and training all users, stakeholders' participation is critical. Stakeholders who have a sense of ownership often create test procedures, implementation plans and training manuals themselves. When testing, they try to go beyond what is written in the instructions to test the features in various operational scenarios, giving feedback for quick fixes and in many cases considering changes in the work processes to work better with the system. Furthermore, when system development is delayed, as it usually is, stakeholders who do not feel ownership typically blame designers. Stakeholders who design their own system try to work out the best solution together.

It is helpful to use Carlile's (2004) distinction of transfer, translation and transformation, as shown in Table 8.2. Although Carlile proposed this distinction in the context of knowledge sharing, it is applicable to information system design. Various levels of design exist: features, work practices, and goals. Features refer to how the information system behaves, e.g., producing a certain output given a certain input. Work practices encompass not only the features but also how people use them to accomplish their work. Goals are high-level objectives that underlie the work practices. For instance, an accounting slip that an accounting department receives and uses to manage their ledger is a feature. The accountants' work practices include looking at the items listed on the slip, approving them, and entering journal titles for each. The goal for this practice indicates that the accounts are responsible for the centralized accounting data and possess control of all accounting activities of the organization. The organization might introduce a new information system to achieve an alternative goal: decentralized and real-time

Table 8.2 Transfer, translation and transformation of design

	Transfer	Translation	Transformation
Created elements	Features	Work practices and features	Goals, work practices and features
Users' actions	Users demand certain features, which are in most cases features of their existing system	Users explore desired work practices by using available features to achieve a fixed goal	Users redefine the goal and design features and work practices
Designers' actions	Designers propose an alternative feature, but users refuse it. Designers cannot easily persuade users	Designers explain an alternative work practice but users are fixated on their goal	Designers refine users' ideas by working out technical details
Boundary of users' work	Users think that designing is not their responsibility but the designers', in exchange for payment	Users understand that they are responsible for design, in particular meeting the budget constraint	Users understand that they are responsible for realizing the value of the technology for the organization
Resulting design	Infeasible and costly	Complex and expensive	Simple and consistent

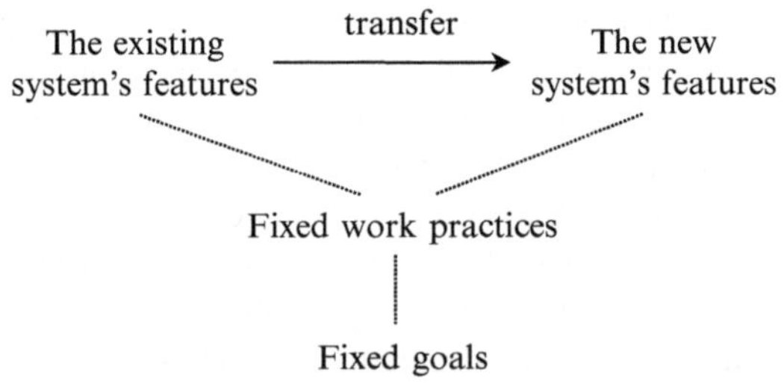

Fig. 8.1 Transfer

accounting for timely decision making. For this goal, all members of the organization, not just accountants, are presumed to enter accounting data and take responsibility for them. Manual approval by accountants for all purchases might no longer make sense, although accountants must still take some responsibility. Furthermore, work practices might require alteration as well as the features.

Transfer refers to an exchange of information out of fixed knowledge (Fig. 8.1). In this case, stakeholders simply state what they want without entertaining the possibility of an alternative perspective. In accounting system development, users might demand a unique identification number for accounting slips, which they had been using in their existing system. After learning that the packaged application did not offer that exact identification number, they simply demanded a modification. The requirement was simply transferred without being altered. Preservation of the same work practice was attempted.

Fig. 8.2 Translation

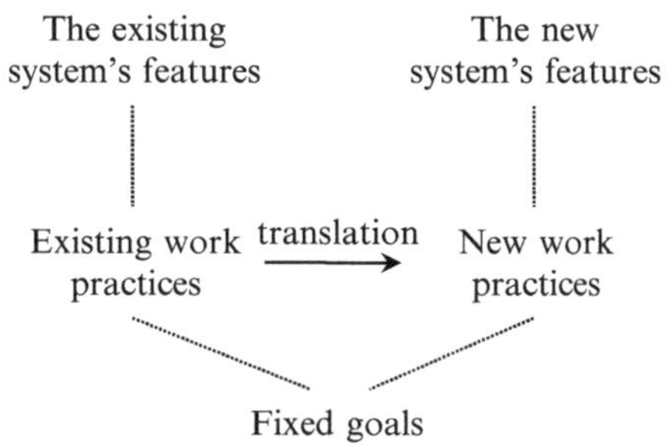

Fig. 8.3 Transformation

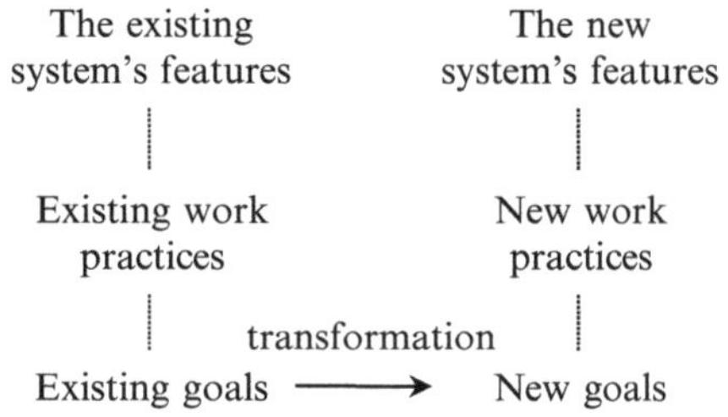

Translation involves negotiation of meanings, although the knowledge is fixed (Fig. 8.2). Stakeholders are ready to change their initial requirements of features and to try to create new features. Typically, these new features aim at the same goal. In the example of the accounting system, users proposed a tweak of the system, e.g., a combination of various fields to realize the equivalent of a unique identification number. Therefore, users proposed different work practices although they tried to achieve the same goal. Although the original requirement was translated into something different, the underlying meaning and the goal remained the same, achieving the unique identification number.

Transformation, however, involves change in the goal for which features are considered (Fig. 8.3). The identification number had been used in the character-based terminals of the existing system, although no equivalent number was exposed to users of the packaged application because users could simply search for slips by numerous fields and click on the result—users need not worry about numbers. The character-based terminals require the accountants to type numbers to look up slips instead of clicking with a mouse. The identification number was also needed to search for paper documents attached to slips such as invoices and receipts. However, because the new system was intended to store scanned documents, no number was needed to look up the documents. Therefore, the whole idea of the identification number might have been altered. In this case, the goal was altered. As a result, work practices and features were altered.

It is not easy for stakeholders who have their own ingrained worldview to learn to design from a new perspective. In many ways, what stakeholders suggest early on is not what they end up needing later. Many features are fixed at the level of transfer or translation. In information system development, stakeholders must learn up to transformation within a limited time. Without this learning, designers often try to arrive at compromises. For instance, translation might be sufficient because the tweaks can be inexpensive, but design might not achieve the holistic goal.

8.3.2 Boundary Objects

Several theorists have reported that boundary objects play a crucial role in this transformation (Bechky 2003; Carlile 2002, 2004). Boundary objects are objects such as artifacts, locations, plans, and concepts that can be used by multiple groups that have different work practices and goals. Boundary objects help represent stakeholders' interests. One party's concern cannot be expressed clearly in words. Nevertheless, boundary objects translate the concern in a form that is readily understandable to other parties. For instance, a certain design diagram of product designers can represent a concern of production engineers. Product designers can readily understand the important issues for production engineers. Similarly, boundary objects help users understand an alternative worldview.

In packaged application implementations such as ERP, users can play with a demo system before discussing detailed requirements such as those for a conference room pilot. Although the package will not be able to support a feature that they want, they learn that they can perform the same task with alternative features that are available in the package. After some time, users start to think of alternative technical implementations. In one project, users started to propose the creation of a "virtual warehouse" to simplify the work process. The same users initially did not have a concept of a virtual warehouse because warehouses for them are always physical ones. By playing with the system and exploring the features, they were able to start to think differently.

When no fully functional system exists for use in the demo, low-fidelity mockups can be useful (Ehn and Kyng 1991). Often mockups are made of paper with no computational capability. Furthermore, it is often more important for users to write design documents rather than to talk about designs. By specifying the detailed design, e.g., workflows and UI, users come to think of the whole system. Yet, because it is not easy for untrained users to specify design, a paper-based toolkit is sometimes used (Muller 1991; Yamauchi 2011). Users can design by moving stickers on a formatted sheet. Various stakeholders collaborate and specify a single design.

8.3.3 Power

Power is critical in relation to users' learning. In traditional participatory design, users are portrayed as powerless. Situated learning theory suggests that access is crucial for learning (Lave and Wenger 1991). Because stakeholders are sequestered from design, their learning is inhibited. This situation has apparently changed over the last decade. Information technology has become a commodity. Users' reliance on IT professionals' expertise has been less critical. Competition has strengthened among IT vendors, who now must compete in terms of price. In this circumstance, stakeholders often have power over IT designers because designers come from an IT vendor that is contracted by the client. Customers have power as makers of purchasing decisions. In development of large information systems, somebody else might decide to purchase a system. For instance, top management or the IT Department often decides to introduce a system. Yet, users must eventually have a say in detailed designs. Many projects fail because of users' resistance. Therefore, designers must obtain agreement from users on detailed features.

Designers often feel powerless in interacting with stakeholders who do not want to participate and instead only dictate the design. Stakeholders demand features which they believe are important. Designers bring several alternative designs but stakeholders simply reject all of them. Although powerlessness hinders learning, power can also be consequential to learning. Powerful stakeholders simply transfer requirements without altering their perspective.

Designers are not powerless, however. Stakeholders depend on them to detail the features. Precisely because stakeholders must pay for the system, designers often suggest that they could simply listen to them but the system will be too costly. In typical information system development, stakeholders' requests inflate the project and cost estimates explode the confines of the budget. By imposing pressure on stakeholders to reduce costs, designers can leverage their power.

Beyond learning, power is always an issue in information system development. Design is always a political process because design affects the life of each stakeholder (Franz and Robey 1984; Markus 1983; Robey et al. 1989). New information systems often alter the power structure of an organization. A group that monopolized centralized information might lose its power because the system aggregates information for use by other departments. Alternatively, the headquarters might try to centralize information and control. Sales departments want to control inventories of products at factory warehouses while factories have managed these inventories. Because factories want to ship goods out of warehouses to recognize revenues, they push the good onto the sales department's ledger. Understandably, those in the sales are frustrated because they receive some goods that they did not request. From the factories' perspective, sales people who do not come to the warehouse cannot manage the inventories because space is limited and much work is involved in maintaining the volume. The affected parties therefore seek to influence the design by resorting to the power of their managers and by building alliances.

Conflict of this kind is observed in almost all system development projects. Some projects can efficiently resolve these conflicts if top management gives top-down directives. Nevertheless, parties seek to maintain their power by manipulating details. Some group might refuse to use the system, thereby dooming the project to failure. Unfortunately, no shortcut exists to this sticky problem. To resolve these conflicts, practitioners must understand the political structure correctly and facilitate a dialog among groups. When we try to involve users in design, we must consider identifying and possibly altering power so that transformation of users' knowledge can be induced.

8.4 Expanding the Scope of Participatory Design

8.4.1 Technology Design in Research

As described above, participatory design is not merely involving stakeholders in design. It is about designing a socio-technical system of which stakeholders' reflexive agency is a part. Next technology design and development in a research setting are discussed, particularly in industry R&D. Because every socio-technical system needs participatory design in a broader sense, we must think of participation in R&D activities. Participatory design has been applied in research settings. Typical participatory design methods are useful in research settings. Particularly when researchers are developing a technology for end-users, understanding these users through observations and exploring design in workshops together is a crucial step. Often, various technologies developed through usual R&D channels are produced in absolute ignorance of the actual users' situations.

Yet, researchers must do more. In a typical research situation, a definite group of stakeholders exists. Researchers are trying to design not a concrete product or service but technologies that have not existed before. For that reason, researchers must consider such issues as integration into the existing technical infrastructure, market adoption, business models, and regulatory and institutional arrangements in addition to technical functionality, performance, feasibility, and reliability. Researchers must therefore deal with the whole socio-technical system.

Before examining details, we must understand the current research environment, particularly in industry research labs. Industry research labs have incorporated consideration of market demands. Inventions and innovations tend to take place where demand exists (Griliches 1957; Nelson and Winter 1982; Schmookler 1966). Nonetheless, researchers in industry labs often do research whose connection to a corporation's businesses is not necessarily readily apparent. Even if demand is considered, it is detached from the researchers' research practices. In traditional R&D organizations, researchers have been doing only research, i.e., designing and conducting experiments, designing and implementing prototypes, and reading and writing papers. That investigation is done either before the research is launched or by managers who do not do the day-to-day research for which demand is considered.

In the current environment, more pressure is added to research labs to produce technologies related to existing businesses and to create new businesses for corporations. This pressure has been transferred to individual researchers. When researchers have an idea they want to work on, they must show that it has market potential. In many cases, marketing professionals and managers determine that no sufficiently large market exists and decide not to fund it. Even if it is funded, after a year or two, when the project is evaluated, it is either reduced or shelved because of unclear market potential. This situation creates two problems. First, research that addresses a concrete demand that already represents a large market tends to be incremental and does not ensure the long-term advantage of the organization. Second, researchers themselves are frustrated.

In such situations, researchers must be conscious of how the technologies they create will be used. They must have a certain understanding of the technological, economic, and social conditions in which the technologies are used.

8.4.2 Integrating Technology and Market

In recent years, it has been witnessed that some researchers take an increasingly proactive role in exploring markets just as they explore science and technology—interacting constantly with users, potential clients, domain experts, competitors, partners, funding agencies, and investors. Researchers at Palo Alto Research Center (PARC), a Xerox company, started a new research program in clean technology, particularly solar energy generation, in 2004. Researchers had no idea what to work on in the solar field. Moreover, they did not know anybody in that domain. They slowly started exploring the domain. For instance, researchers formed a study group to read papers and books, invited prominent speakers and had meetings with some players in the domain. They also went to the largest conference in the domain, the World Renewable Energy Congress, and listened to as many talks as possible. They met several key figures in the domain and returned with a stack of business cards.

After concluding the solar concentrator photovoltaic (CPV) research, researchers at PARC wanted to work on crystalline silicon panels. Nevertheless, it was a huge area and they had no idea what to work on specifically. One idea was to start experimenting with new materials, methods, and systems for PV panels. Instead, they explored the domain by talking to various domain experts to identify something more realistic. When they were observing the manufacturing line, they encountered something familiar: a printer. As a research center of Xerox, they knew about printers better than anyone. They learned that screen printers were used to deposit gridlines, metal lines to extract electricity from PV. Ideas for improving this line came easily. They thought about printing taller and narrower lines. The line is put on top of the cell. Therefore, the shadow it creates raises inefficiency. The taller line was able to produce a smaller shadow and reflect some light on the cell. For this, they invented co-extrusion printing.

Yet, researchers did not start investing money in this research, choosing instead to be cautious. They decided to do two things. First, they hired a domain expert who had more than 30 years of experience in solar research and business areas. They worked with this expert to explore new ideas and refined designs. This expert, a former executive of a large PV manufacturer, also guided the researchers in business decisions. He was able to evaluate an idea quickly and offer some advice based on his experience. Second, with this expert, researchers approached and consulted with several potential clients. They explained the technology's performance, reliability, expected lifetime, and targeted price. Positive feedback brought them more confidence. Only after this validation did they start to invest in this research and develop the technology.

This case shows that researchers explored not only the technology but also the market demand, potential client requirements and investment criteria, existing technological infrastructure, and business models. The kind of market understanding they developed was much more detailed than the casual market sizing that scientists typically do to justify their project. Business analysis revealed that simply developing the co-extrusion printers did not justify the investment. They quickly realized that the business would depend on selling ink, which shaped the research direction early on. The printer needed to be designed in such a way as to operate within the existing technological infrastructure. More than anything, the need for this technology was validated. Co-extrusion printing offered performance advantages that were sufficiently large to make a difference in overall module efficiency. In addition, this printing used a non-contact method: the printhead does not touch the wafer. The price per watt was the most critical factor for solar power. Therefore, PV manufacturers came to use very thin wafers. The non-contact method was appealing because it will not break the wafer.

The researchers explored new research themes not by sitting in the lab and reading scientific articles but by interacting directly with the market. When researchers started biomedical research, another new domain for PARC, they thought that inkjet printing technology could be used for DNA microarrays. The idea was excellent and an important technology was developed: high-volume production disposable print heads. Yet, the technology had no outlet because DNA microarrays were not used in clinical settings but mostly in research, for which high volume production was unnecessary. This lesson made the researchers take greater care in choosing a research theme.

In the biomedical case, researchers were able to identify domain experts who were able to guide them. They were connected to faculty members of Scripps Research Institute and formed a joint research institute, the Scripps-PARC Research Institute. PARC researchers asked Scripps researchers for a list of the ten largest problems in the field. PARC researchers explored these problems and found several that were particularly interesting because they possessed the capabilities to address them. One, for instance, was detection of cancer cells in blood. This problem was difficult: no viable solution had existed because cancer cells are so few in blood. PARC researchers realized that this problem was similar to laser scanning in printing. They thought about using laser to illuminate tagged cells and collect light with a sensor. Unlike laser printing, again based on the long history of research for Xerox, a challenge was to collect the light at high speed. One researcher came up with the idea of using a bundle of optical fibers to channel the light into a sensor. Thereby, an extremely fast and accurate detection system was invented. In this case, too, scientists worked closely with various domain experts including physicians.

8.4.3 *Iterative Exploration*

Research is fundamentally uncertain. No matter how well researchers anticipate possible contingencies, accurate prediction of the future is not possible. The best approach appears to be an iterative one: Testing one's understanding and adjusting the course of action frequently. Although it is a standard practice to organize product development concurrently, a serial process appears to be more appropriate for uncertain research, i.e., tackling one thing at one time.

A group of researchers at PARC invented a new way to filter water. Traditionally, membrane-based filtration or chemically processed sedimentation has been used to remove particles from water. Researchers who had formerly worked on toner particles invented a novel approach that required no membrane and achieved high throughput. The idea was to move fluid through a curved channel. Particles move to either the inside or outside of the channel depending on the design, thereby forming a band. A separator at the end can then remove the band. The innovation of this technology was the use of Dean's vortex in addition to centrifugal force. Even particles with neutral buoyancy can be moved to a side through the vortex created in the channel. Because this technology requires no membrane, less maintenance is necessary, little energy is used, and throughput is high.

After these researchers developed a design through experimentation and simulation for 3 months, all of them stopped the research and started doing market investigation. They wanted to know where this technology could be used. Filtration is quite versatile. It is useful not only in municipal water treatment but also in grey water, blood, desalination, and even wine production. It took them 3 months to have a good understanding of possible applications. After they identified possible applications and understood basic requirements for each, they went back to research to refine the technology for targeted applications.

The notion of real options can be useful. The future cannot be fully determined. Therefore, researchers justify a small investment to gain more information in the future. The idea is not to predict the future accurately. Researchers must identify and address the most critical issue and adjust their commercialization scenario as they learn more. Although technical risks demand extensive research and therefore investment, it does not mean that researchers must dive deep into doing the research and show the result only a year later. They can specifically examine it if they identify the most risky component of the technology and devise steps to address the risk.

The risk can also be economic or social. For instance, user adoption is the greatest uncertainty in most cases. Not all good technologies will sell well. It is not possible to come to a determinate answer of whether a particular technology will sell. Yet, many risks can be reduced if one talks to potential clients and understands the economic ecosystem. The most critical issue might not be resolved with a small investment. In this case, researchers must manage the risk rather than reduce it. They must think of the smallest step to gain information for reducing the risk. For example, the researchers who invented the filtration technology were aware that the largest market was lucrative but risk-averse. They visited a water plant and realized that it would take much longer to address this market. They changed their strategy and explored another market.

Through the iteration, researchers can also find unintended applications. The filtration researchers, for instance, discovered that treatment of water used in oil extraction was a critical problem that the technology was able to address. This application was not obvious to the researchers who did not know the domain. Only by reaching out to stakeholders in the market were they able identify that application. In this sense, market understanding also evolves as the technology evolves.

Researchers often come to terminate or shelve their project by themselves if they go through this learning. For instance, as researchers at PARC went through the iterative process, they found themselves unable to resolve a critical issue related to the market structure, which led them to shelve the project and propose a new research project, based on what they learned. Similarly, in a solar concentrator project, researchers approached potential clients before investing in the project. Because they received mixed feedback, they changed the commercialization strategy and quickly licensed the technology without a substantial investment. Again, this experience led them to propose a new project to avoid the problem they had encountered in the previous project.

In sum, any technology development in research must be regarded as a design of a socio-technical system. Because stakeholders' reflexive understanding is part of the equation, researchers cannot design a technology they conceive in their lab and put it out in the world. They must inquire, test, and refine the technology repeatedly. In such research, there is seldom a definite group of users with whom to interact. Researchers must involve potential clients, partners, domain experts, and competitors as well as dealing with regulatory and institutional factors. The iterative development of the technology and the market in which it will be used is key to successful research and development.

8.4.4 Continuity and Discontinuity

It is often suggested that researchers in R&D must take risks and engage in disruptive research. In other words, discontinuity is encouraged. Yet, we must also think about the socio-technical systems in which research activities are embedded. As described above, PARC researchers moved into completely new domains such as those of clean technology and biomedical technology. Apparently, they conducted discontinuous research. A closer look reveals that researchers carefully sought and constructed continuity in moving into discontinuous domains.

The challenge for researchers was a lack of credibility in the new domain. Their managers were not comfortable taking risks. People in the new domain also did not provide support readily. It was only through continuous means that researchers were able to overcome the challenge. For instance, as we saw, they looked for problems for which their existing capability could be leveraged: printing for solar, laser scanning for cancer cell detection, and particle manipulation based on toner research for water filtration.

Researchers also relied on external domain experts. These experts played a crucial role not only in filling the gap in knowledge but also to legitimate the researchers. Managers would feel more comfortable if external domain experts were able to endorse the researchers' ideas. Researchers could also gain access to networks that the experts had in the new domain. PARC institutionalized involvement of these experts by creating a new title: *visiting technologist*. Visiting technologists stay at PARC for a certain length of time. Currently at PARC, most new research projects hire visiting technologists.

Participatory design in a broad sense must encompass the socio-technical system of research itself as well as that of the technology. Research must fit into the institutional environment. Although discontinuity is encouraged in the abstract and achieved in retrospect, the research process must be continuous.

8.5 Summary

This chapter presented the argument that participatory design is necessary not only to incorporate users' voices in design but also to conceive the overall socio-technical system of which stakeholders are a part. The commonsense notion of design, rooted in artifact design, does not represent what this design really is. The critical issue of stakeholders' learning was discussed. Subsequently, the scope of participatory design was expanded in the context of industry R&D, in which technologies are developed iteratively in conjunction with the users, customers, investors and other players in the market.

References

Bechky, B.A. *Sharing Meaning Across Occupational Communities*: The Transformation of Understanding on a Production Floor. Organization Science 14(3) 312–330, 2003

Berger, P.L., Luckmann, T. *The Social Construction of Reality*. Doubleday, Garden City, New York, 1966

Bodker, K., Kensing, F., Simonsen, J. *Participatory IT design: Designing for Business and Workplace Realities*. MIT Press, Cambridge, Massachusetts, 2004

Carlile, P.R. *A Pragmatic View of Knowledge and Boundaries: Boundary Objects in New Product Development*. Organization Science 13(4) 442–455, 2002

Carlile, P.R. *Transferring, Translating, and Transforming: An Integrative Framework for Managing Knowledge Across Boundaries*. Organization Science 15(5) 555–568, 2004

Ehn, P., Kyng, M. *Cardboard Computers: Mocking-it-up or Hands-on the Future*. In J.M. Greenbaum and M. Kyng (eds.), Design at Work: Cooperative Design of Computer Systems. L. Erlbaum Associates, Hillsdale, New Jersey, pp. 167–195, 1991

Franz, C.R., Robey, D. *An Investigation of User-Led System Design: Rational and Political Perspectives*. Communications of the ACM 27(12) 1202–1209, 1984

Giddens, A. *The Constitution of Society: Outline of the Theory of Structuration*. University of California Press, Berkeley, California, 1984

Greenbaum, J.M., Kyng, M. *Design at Work: Cooperative Design of Computer Systems*. L. Erlbaum Associates, Hillsdale, New Jersey, 1991

Griliches, Z. Hybrid Corn: *An Exploration in the Economics of Technological Change*. Econometrica 25(4) 501–522, 1957

Kuhn, T.S. *The Structure of Scientific Revolutions*. University of Chicago Press, Chicago, 1962

Langdon, W. *Do Artifacts Have Politics?* Daedalus, 109(1) 121–136, 1980

Lave, J., Wenger, E. *Situated Learning: Legitimate Peripheral Participation*. Cambridge University Press, Cambridge, 1991

Markus, M.L. *Power, Politics, and MIS Implementation*. Communications of the ACM 26(6) 430–444, 1983

Muller, M.J. *PICTIVE — An Exploration in Participatory Design*. CHI 1991, Proc. SIGCHI Conference on Human Factors in Computing Systems: Reaching Through Technology, 1991

Nelson, R.R., Winter, S.G. *An Evolutionary Theory of Economic Change*. Belknap Press of Harvard University Press, Cambridge, Massachusetts, 1982

Robey, D., Farrow, D.L., Franz, C.R. *Group Process and Conflict in System Development*. Management Science 35(10) 1172–1191, 1989

Schmookler, J. *Invention and Economic Growth*. Harvard University Press, Cambridge, Massachusetts, 1966

Schön, D.A. *The Reflective Practitioner: How Professionals Think in Action*. Basic Books, New York, 1983

Schuler, D., Namioka, A. *Participatory Design: Principles and Practices*. L. Erlbaum Associates, Hillsdale, New Jersey, 1993

Simon, H.A. *The Sciences of the Artificial*. MIT Press, Cambridge, Massachusetts, 1968

Trist, E.L., Bamforth, K.W. *Some social and psychological consequences of the longwall method of coal-getting: An examination of the psychological situation and defences of a work group in relation to the social structure and technological content of the work system*. Human Relations 4(1) 3–38, 1951

Yamauchi, Y. *Transforming Information System Design: Enabling Users to Design*. In M.H. Szymanski and J. Whalen (eds.), Making Work Visible: Ethnographically Grounded Case Studies of Work Practice, Cambridge University Press, New York, 2011

Yamauchi, Y., Swanson, E.B. *Local assimilation of an enterprise system: Situated learning by means of familiarity pockets*. Information and Organization 20(3–4) 187–206, 2010

Chapter 9
Case Writing

Shigeo Matsubara

Abstract This chapter describes case writing as a method of conveying knowledge in the field. "Standing on the shoulders of giants" is a famous utterance by Sir Isaac Newton. It has also been adapted as a phrase that appears at the top of the page on Google Scholar, which provides a search service for scholarly literature. This quote expresses the idea that no matter how great one's achievements, success is always based on the research and work of thinkers from the past. As it is vital in research to accumulate past work, researchers need to make their findings accessible to others. This can be done by publishing papers in scholarly journals. However, if researchers want to publish various lessons obtained from practice in the field, they are often faced with difficulties in explaining why the lessons are significant. Thus, developing methods of conveying knowledge in the field remains a challenge. Further, knowledge is not always communicated from one researcher to another; it can also be passed from researchers to practitioners. There have recently been increasingly more areas—e.g., the design of information systems—that have required coordination between theoretical studies and activities in the field. This means that researchers are faced with the challenges of how to communicate research findings to others and how to develop practitioners. Here, the important thing is not simply knowing information, but also acquiring knowledge that can actually be applied. Be it as a designer or a user, one must confront the field and face these issues when handling information. To help resolve these two problems, this chapter focuses on cases and outlines the methodology for case writing, using the introduction of a new system in the workplace as an example.

S. Matsubara (✉)
Department of Social Informatics, Kyoto University, Yoshida-Honmachi, Sakyo-ku,
Kyoto 606-8501, Japan
e-mail: matsubara@i.kyoto-u.ac.jp

T. Ishida (ed.), *Field Informatics*, DOI 10.1007/978-3-642-29006-0_9,

© Springer-Verlag Berlin Heidelberg 2012

9.1 What Are Cases?

Case writing is the act of composing a case to convey some knowledge to resolve a research question where the phenomena and the contexts are difficult to separate. Here, a case is a description about a real-life event in a specific circumstance.

With this definition, it may seem that almost all descriptions of real-life events could be considered to be cases. Many people have recently used blogs or social networking services (SNS) to write accounts of their travels. Although tales about the difficulty of communicating in foreign languages or descriptions of food in restaurants must be descriptions of specific events that have actually occurred in unique environments, cases are meant to involve the writer's clear objectives as to whether they are for research or for teaching purposes.

Figure 9.1 lists various case classifications. As previously mentioned, this chapter discusses ways of conveying field-related knowledge, i.e., communication of academic knowledge from one researcher to another, and communication of practical knowledge from researchers to practitioners. The former is called case studies for research purposes, while the latter is called case studies for teaching purposes. A case can be viewed as the outcome of a case study for research purposes, while a case can also be viewed as the course material in a case study for teaching purposes. Some cases involve a mixture of both, but one should treat the two separately when writing a case.

The term "case study" has a wide variety of meanings: It can refer to a research method or its outcomes, or a teaching method using cases, which often causes confusion. Note that a "case study" will mean a research method throughout the rest of this chapter. Various research methods are introduced in this book. Even if we focus on the process where a researcher makes and verifies a hypothesis to examine a research issue, there are various research methods such as logical proofs based on mathematics, simulations, experiments with participants, and surveys administered using questionnaires. Case studies, which stand alongside these methods, are also used to examine and analyze particular cases, verify hypotheses and derive general truths from the results. Cases describe the process of going back and forth between hypotheses and analyses, giving readers perspectives on an author's thought processes and research strategies.

However, the Harvard Business School provides many examples of well known cases for teaching purposes. The case method, developed during the 1930s at Harvard Business School, is now used by many business institutions. It is not only applied to management, but also to the legal and medical fields. Much theoretical knowledge on finance and accounting is available in management and other fields, but these theories are often difficult to apply to actual situations that occur in society, as their underlying causes are often too complicated. In such cases, the challenge is to develop the ability to tie theories to practical situations. Such knowledge has to be obtained by students themselves because it is difficult for them to obtain this only by attending lectures (Garvin 2003). The idea, then, is that presenting a case as an apparatus for inducing debate about the "why" and "how" of a specific case can help to create the practical knowledge required in actual field work.

a) Outcome of the case study (research purposes)
- This is used to verify the hypothesis
- Complete and accurate descriptions of a real event are required
- Conclusions by the author are needed

b) Cases used in the case method (teaching purposes)
- Apparatus for inducing a debate
- Whether it brings about an active debate is more important than the completeness or accuracy of the descriptions
- Conclusions by the author are not needed

Fig. 9.1 Classification of cases

Accurate descriptions of real events are not required in cases for teaching purposes compared to those for research purposes. Instead, the main focus in cases for teaching purposes is to set a framework that will enable discussion among students. Herein lies a significant difference. Teaching-purpose cases are essentially less constrained from the viewpoint of case writing, while the lecturer's skill of facilitating a discussion, as well as the content of the cases, is important. Thus, we will focus on case studies and provide supplementary information about the case method when necessary.

If we consider the format of cases, the main difference between cases for research purposes and those for teaching purposes is whether authors have to write conclusions. Obviously, conclusions by authors are needed in cases for research purposes. That means that readers of research-purpose cases can acquire insights into how and why a certain issue exists. Teaching-purpose cases, however, force readers to think about the decisions they would have to make to solve a given problem.

In the remainder of this chapter, Sect. 9.2 presents an example of a case, Sects. 9.3, 9.4, 9.5, and 9.6 explain when and how to use case studies and case methods mainly based on discussions by Yin (2008), and Sect. 9.7 discusses cases in field informatics. If readers are already familiar with case studies and case methods, and interested in the relation between cases and field informatics, move on to Sect. 9.7.

9.2 An Example of a Case

This section illustrates the analysis of a case included in the case studies by presenting a scenario where accounting software is upgraded in a company. This is a fictitious case invented by the author after consulting revised materials on tax accounting software. The purpose of this section is to give readers an intuitive understanding of case descriptions. To complete the following case, the author would need to add information pertaining to the process of going back and forth between hypothesis and analysis, as well as presenting his/her conclusions.

Case: Accounting System Upgrade

Mr. Kojima is an assistant manager for the hardware development department at A Telecommunications. He received an email from the main accounting department asking him to attend an informational meeting about the launch of an operations management system.

The Financial Services Agency's Business Accounting Council released a written opinion (1998) and practical guidelines (2000) on software. With the new guidelines, previously unclear software accounting procedures were defined, and tax laws were revised. Thus far, tax authorities have issued instructions on software development for internal company use that outsourcing expenses must be capitalized and internal costs are permitted to be processed as period costs. Commission expenses must be capitalized in terms of accounting. Under the new standards, software production costs, calculated based on appropriate cost accounting, must be processed as intangible fixed assets (with 5 years as the normal service life) in both the accounting and tax rules. If the existing package software was basically processed in the IT department to meet company specifications, labor costs and expenses in the IT department would have to be capitalized as "software development expenses." Thus, the company is now forced to tally software developer work hours and calculate the cost price using labor costs per hour and expense ratios. To comply with these standards, the company's accounting department decided to develop and implement a system to calculate the number of software developer man hours.

The company held occasional informational meetings about the improvements to its accounting system. These sessions often consisted of nothing more than cursory explanations, and Mr. Kojima was unenthusiastic about attending them. The development departments for hardware and software at A Telecommunications are scattered in various locations, and informational meetings are mostly conducted via teleconferencing, which helps to save on travel expenses. As long as one was there gathered around the monitor, he/she could bring in other work and the other participants would have no idea whether he/she was taking notes or working on other tasks. That made the meetings even less exciting. At the meeting, Mr. Kojima sat down next to the manager, Mr. Takemura, who was head of Development Department 1. Mr. Takemura asked Mr. Kojima, "Did you bring something to keep you busy? We are going to have a lot of time to kill." Mr. Kojima replied, "Yeah, I guess." As Mr. Kojima could not simply ignore the discussion at the meeting because of the nature of his position, he decided to tune in. However, as he was often unable to provide answers to questions about the accounting system from members of the hardware department, inquiries were frequently directed to the main accounting system manager. Therefore, many members directly contacted the main accounting department.

The head of accounting appeared at the meeting and only told the participants that the subject of the session was very important. After that, the accounting system development manager began his explanation. Mr. Kojima respected the head of accounting for letting himself be pulled away from more important work just to give a short greeting. As expected, the meeting simply stuck to the script of the distributed materials, and there was very little information worth writing down. Mr. Kojima always felt that if all he had to do was look over the materials, he could have finished up in 5 min—there had to be a more efficient way of holding informational gatherings, since workers had to look over the materials again to figure out how to use the software when they actually had work in front of them.

As the meeting went on, Mr. Kojima heard that every development department member had to enter their software development man hours into the system. If no section members were involved in software development, however, all one would have to do is just complete final processing of registration. Mr. Kojima did not belong to the section, but rather the development department's management administrative staff. As there were not many people on the administrative staff, they never developed any software, except for outside orders. Thus, Kojima did not think that the launch of a new operations management system would create a huge workload.

Mr. Takemura's Development Department 1 included software developers. If a department had a software developer, the system specifications required all members of the corresponding section to enter the number of hours they had worked on software development. Mr. Takemura understood that the new system would prevent people from forgetting to enter work times or entering incorrect figures, but he did not think it was necessary to force people not involved in software development to input a "0" for development time every single day. System development follows a four-step pattern: (1) analysis of current workflow, (2) basic design, (3) detailed design, and (4) programming. Steps (1) and (2), which fall under the category of advance preparation, are excluded from production costs. Lost time caused by design changes and other factors is also excluded. Mr. Takemura knew that when he informed section members of these changes, they would ask him if "these hours should be included in the operating time." He then decided to leave the input tasks to his assistant, Ms. Kadowaki.

Ever since the former manager, Mr. Yamazaki, who was Mr. Takemura's predecessor, had been in charge of Development Department 1, most clerical tasks had been assigned to assistants. Headquarters had taken an inconsistent stance on hiring assistants, at one point telling departments to cut the number of employees to cut costs, but at another asking departments to make active use of a subsidiary temporary staffing agency. The previous manager, Mr. Yamazaki, left many tasks to these assistants, from filing forms and performing other traditional duties to managing work schedules and taking care of orders for various articles. The most recent manager, Mr. Takemura, had merely followed suit.

Mr. Takemura decided to have section members send emails to Ms. Kadowaki with their monthly software development work times, which Ms. Kadowaki would then input into the system. Mr. Takemura thought that Ms. Kadowaki would also be able to prod employees who had yet to report their times under such a system.

Near the end of the meeting, the floor was opened up for questions. There were no questions from the teleconferencing room where Messrs. Kojima and Takemura were stationed. Mr. Takemura asked Mr. Kojima what he thought of his plan to have Ms. Kadowaki handle the input work. Mr. Kojima could not see any real problems with the plan, and even if something were to happen in the future, he knew in the back of his mind that he would already have moved on to another department, so he told Mr. Takemura it was a good idea. Assistant managers at A Telecommunications were essentially caretakers and if they were ordered to be assistant managers they waited around for different postings. Thus, Mr. Kojima did not give much thought to the matter.

After the informational meeting, Mr. Takemura returned to his room at Development Department 1 and explained the new job to Ms. Kadowaki who had never thought she could go against instructions from her superior, but since the task was a relatively simple one, she accepted the assignment. Still, she was puzzled. As accounting system accounts are not issued to assistants, Ms. Kadowaki would have to borrow a section member's account to do the work. If she ran into a problem in the system, she would then need to call the main accounting system department at headquarters to ask for help. However, she thought headquarters might be suspicious that a temporary employee like her was involved in work that went beyond the scope of regular duties for assistants.

9.3 Subjects of Case Studies and Case Methods

Although the method of case studies can deal with various subjects, it is not good at everything. The research methods include surveys and experiments as well as case studies and different methodologies are suitable for different subjects. It is crucial to understand this and choose a methodology to fit the situation. We will compare the case studies method with others in this section and illustrate examples of subjects suitable for the case study approach.

We will first describe the conclusions. Case studies have an advantage over other methods if the types of research questions are "how" and/or "why", and the boundaries between phenomena and contexts are not obvious. We will elaborate a little on this.

Yin compared case studies with other research strategies such as surveys, experiments, archival analyses, and history (Yin 2008). He discussed these strategies in the social sciences. Here, we will try to discuss research strategies from the viewpoint of field informatics. Thus, we will compare case studies with surveys, experiments, and simulations. As Yin had done, we will illustrate the characteristics of case studies by comparing case studies to other research strategies in terms of the form of research questions and whether we need to control behavioral events. Table 9.1 summarizes the comparisons of four research strategies.

Surveys including questionnaires are often used for field research, which focus on research questions that ask "who", "what", "where", "how many" and "how much". For example, in the accounting system upgrade, a survey would have been an effective way of finding out if respondents felt that participation in informational meetings would be beneficial or thought that shorter/longer meetings were better.

Table 9.1 Comparisons of four research strategies

		Surveys	Experiments	Simulations	Case studies
The form of research questions	Who	✓			
	What	✓			
	Where	✓			
	How many	✓			
	How much	✓			
	How		✓	✓	✓
	Why		✓	✓	✓
Control of behavioral events		Not needed	Needed	Needed	Not needed

This method does not have to control the behavioral events. On the contrary, if the procedure for sampling is biased, the findings from surveys are no longer reliable. The drawbacks of this method are that carrying out in-depth discussions is difficult and that large-scale surveys incur large costs. A cross-sectional investigation into the information available to employees when a new system is being introduced would be an expensive venture.

Experiments should focus on the research questions of "how" and "why". Returning to our accounting system upgrade scenario, a good example of this method would be an attempt to examine how prior explanations about the system can reduce mistakes in subsequent uses of it. This method requires behavioral events to be controlled. Bias by participants in the experiment should be removed, otherwise the participants' habits in actions, e.g., whether they were well-organized or lazy, would affect the results. A preliminary experiment might be carried out to remove such bias. Also, as handling the previously discussed accounting system (as it is in the experiment environment) is a complicated process, one might consider replacing it with a simpler system for investigations. If the research questions could be affected by several factors, comparative experiments on all factors should be carried out, which would incur larger cost. Time and budget are often limited, thus only characteristic factors should be investigated. Researchers must also sufficiently address the question of whether behavior that happens in the laboratory actually occurs in the real world. For instance, one must consider whether participants have sufficient motivation to take part in the experiment.

Simulations should also focus on the research questions of "how" and "why". This method also requires behavioral events to be controlled. Research issues investigated through simulations are raised by actual events, but researchers are more concerned about the future, e.g., carrying out what-if analyses. In our accounting system upgrade example, suppose that we had looked at how the behavior of others influenced the accuracy of data input. Here, by creating a decision-making model for each section member and conducting simulation in which the level of the effect of others was changed, one could evaluate the accuracy of overall data input. While building a simulator might be costly, once one had been built, the behaviors under various conditions could be investigated inexpensively.

A great deal of flexibility in simulation is an advantage but one criticism is whether the findings reflect actual behaviors in the real world. We need to remember to give sufficient consideration to this.

Case studies should additionally focus on the research questions of "how" and "why". However, behavioral events are not controlled unlike those in experiments and simulations. Large costs are incurred if participants are observed for long periods in case studies. In addition, whether researchers can interview the people concerned and whether they are accepted by local companies or communities should be considered.

As previously mentioned, there are three methods of carrying out experiments, simulations, and case studies on the research questions of "how" and "why". Case studies differ from experiments and simulations in that behavioral events are not controlled. That is, case studies have an advantage over the other methods if the boundaries between phenomena and contexts are not obvious. The fact that the behavioral events are controlled in experiments and simulations means that phenomena and contexts can be kept separate even if this is a trial and error process. The fact that phenomena and contexts can be separated means that independent variables (input) and dependent variables (result) can be determined. If the relations between factors are complicated, on the other hand, and separating phenomena from contexts (separating agents from environments) is difficult, case studies can be a dominant research strategy.

The case in Sect. 9.2 involves a great many factors, such as the organizational structure of the workplace, prior use of the accounting system, and the clerical skills of the members. There are also many other factors, some of which, like company rules, are clearly stated, but others, such as personal relationships within the section, are not. In this example, the phenomena and contexts are multilayered, making them difficult to separate.

As previously mentioned, case studies are a compelling method of research. However, taking the case study approach presents various difficulties. Cases are easy to read even if readers have been insufficiently trained because they do not contain complicated mathematical formulae but only stories. This creates the feeling that anyone can compose a case, but composing an excellent case is not an easy task. Some tools are easier to obtain if surveys are administered, such as those in sampling and statistical testing methods, compared to carrying out case studies. That is, glowing reports can be produced if prescribed procedures are followed. This does not mean that all reports based on surveys are socially and academically significant but surveys are well-developed methods of research. However, case studies are not yet sufficiently well organized in terms of available tools.

Note that cases are not simple descriptions of actual events happening in the field. Cases should include a clear awareness of the issues involved. In the example in Sect. 9.2, the success of introducing the new system depended not only on how easy it was to use, but also on the users' awareness of the task—in other words, how accurately they attempted to carry out processing. The case was written with the objective of finding out how effective informational meetings were in increasing that awareness. Of course, it is permissible to involve an exploratory process in which one first observes a field and then proceeds to approach the problem. In the end, however, what separates a case from a simple description of an example is a hypothesis that serves as the subject of discussion.

In Fig. 9.1, we have distinguished cases for research purposes (case studies) and those for teaching purposes (case methods). For the former, we compared case studies with surveys, experiments, and simulations. For the latter, however, we will compare the case method with lectures. The objective of communication in a lecture is to pass on specialized knowledge, which can be systematized or verbalized. The case method, meanwhile, conveys practical power—insights and integration—which is difficult to systematize and verbalize. Essentially, this is an example of a subject where the phenomena and contexts are difficult to separate.

9.4 Designing Case Studies and Case Methods

We will take a look at specific elements of the case writing process in this section. Figure 9.2 outlines the entire process from case study design to case writing. This section discusses the design of case studies to enable advance preparation. Subsequent case writing is covered in Sect. 9.5.

Discussion on how to write a good paper is not limited to how to organize sections or how to select a writing style but it can eventually be reduced to how to select an appropriate theme. The same holds for case studies. If a researcher wants to compose a laudable case, a superior research theme should be selected. Composing an outstanding case is difficult if data collection and attempts to compose it proceed before clarifying the issues. The research issues should be clarified in advance and the whole procedure for case studies should be looked over.

Let us examine steps 1–3 in Fig. 9.2 more closely. The topic should be selected and research questions should be clarified in step 1. In this step, researchers are aware of issues at the beginning and they may then find appropriate fields to discuss the issues, or may participate in activities at the beginning and then extract research issues. Next, the research questions need to be prioritized, considering the amount of time required to collect data. If the research questions to be tackled are not determined, this causes behavior where too much data are collected, which results in failure to organize or utilize huge amounts of collected data, or failure to collect the necessary data to construct the discussion. In the example of the accounting system upgrade in Sect. 9.2, if we assume that the success of the new system depends not only on how easy it is to use, but also on the users' awareness of the task—in other words, how accurately they attempt to perform processing—the problem centers on what measures have been taken to increase that awareness, whether those measures are sufficient, and possible ways of making improvements.

In step 2, the units of analysis should be clarified. In the example in Sect. 9.2, the subject for analysis was prior informational meetings, and how effective they were was the focus of the investigation. It would also be possible to analyze how the task

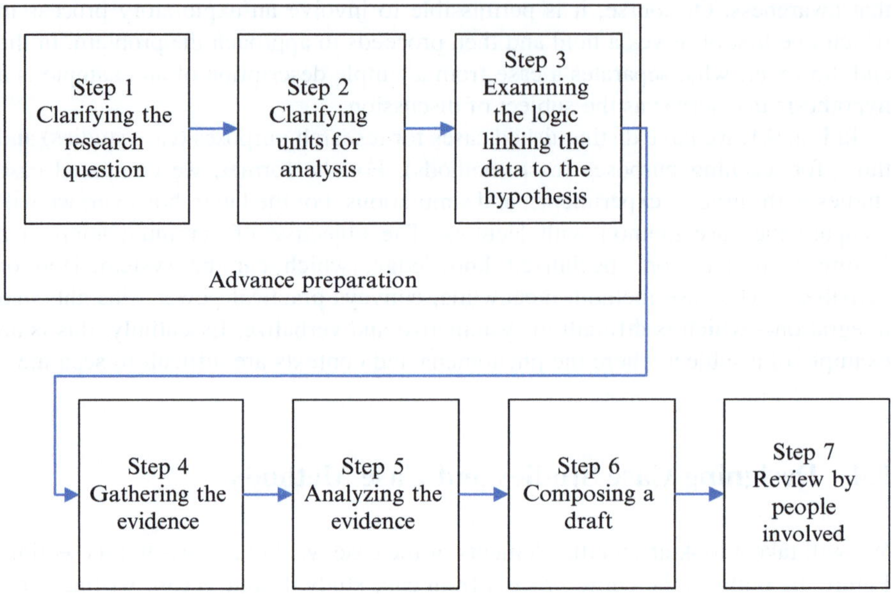

Fig. 9.2 Entire process from design of case study to case writing

processing method of one department and that of an adjacent department influenced each other. Thinking about how to generalize research results is useful in articulating units of analysis.

The logic linking the data to the hypothesis and the criteria for interpreting the findings should be examined in step 3. A method of pattern matching is proposed as the logic linking the data to the hypothesis. This is to describe two potential patterns and to examine which pattern can appropriately explain the obtained data. In the example in Sect. 9.2, evidence was gathered mostly through interviews. However, these avoided directly asking respondents whether the informational meetings had affected awareness. Instead, the interviews had attempted to comprehend changes in awareness by taking into consideration the processing flow in Development Department 1 and its role in implementation.

Now, let us consider the role of theory. The researcher in a case study must understand what the subject of the research is before heading out into the field. In the process of collecting, analyzing, and interpreting observation results, theory provides a road map for the researcher. In the example in Sect. 9.2, it would have been possible to invoke the concept of signaling in game theory. The idea of signaling is that the content of information conveyed at the meeting is not important. Rather, the participants would have recognized how the meeting was important due to the fact that the company had invested the limited work hours of managers to the meeting and had called the busy head of accounting to the meeting just to make a brief speech.

A method of theoretical sampling can be used in the process of deriving a general theory from case studies. Surveys and experiments are based on statistical methods and become reliable if the readers understand the relation between population and samples. This is called probabilistic sampling. Cases, however, are not chosen randomly, but rather those that support the same theory are selected and the theory is then generalized. This is called theoretical sampling. Yin uses the terms of statistical generalization and analytical generalization instead of probabilistic sampling and theoretical sampling (2008). In terms of our example of upgrading the accounting system, this would mean creating an accounting system upgrade case for a different department and investigating whether the department had taken the same problem-solving steps.

Case study design is evaluated in terms of construct validity, internal validity, external validity, and reliability. These in turn are used to evaluate validity at the evidence collection stage, validity at the evidence analysis stage, validity when generalizing results, and validity in the reproducibility of results. There are often concerns about reproducibility with the method of case studies. To address these concerns, it is necessary to record and manage details of the case procedure and ensure that implementing the same case a second time produces the same results.

Many of the points we have discussed thus far would be invaluable not only for case studies, but other research approaches as well. Case studies make it difficult to produce clear results by controlling the conditions unlike experiments. However, a variety of data can be collected in case studies. Thus, case studies become compelling by finding logical links between data. Before research begins, it is important to examine whether case studies are designed so that they fit into this linking of logic.

We need to recall that we distinguished cases for research purposes (case studies) and those for teaching purposes (case methods) in Fig. 9.1. The protocol for the case method was the same as that for the case study. However, one important difference is that analysis is not the responsibility of the author, but rather the students in the class. In step 1, students are shown the type of knowledge they will obtain. The question is sometimes determined beforehand, while in other cases the subject of research has already been determined; either way, it is important to define the intended knowledge. In step 2, details on step 1 are provided as students are encouraged to think about how to generalize a specific case. Then, in step 3, data involved in the case are presented to lead the students to an appropriate discussion. In this step, the instructor moves beyond the case itself and thinks about setting up questions and instructor's notes.

9.5 Composing Cases

This section describes how to compose and write cases in case studies and case methods. This corresponds to steps 4–7 in the case study flow outlined in Fig. 9.2. Some readers may think that "case writing" is an editorial process spanning the rough draft in step 6 and the review in step 7, but "case writing" here includes evidence gathering and other procedures in step 4. As the research subject determines usable data and other factors, it is difficult to make generalized statements about specific things to write about. We will describe some points to bear in mind when writing a case.

What is important is to get rid of any preconceived ideas. Researchers should not bias their thinking toward presupposed results when they collect the evidence or compose cases. Drawing conclusions beforehand and collecting materials arbitrarily to avoid any contradictions is not research. If they happen to obtain information from an interview that contradicts their hypotheses, for example, this should not be thought of as failure, but rather it should be welcomed because it offers possible clues toward new discoveries. Methods of preventing researchers' thoughts from becoming biased are developing case study protocols and building case databases.

A case study protocol includes the specifications for procedures and general rules as well as methods such as investigating the literature or conducting interviews. For example, the protocol includes information about who should conduct interviews and what questions should be asked. As previously explained on the subject of reliability, this is preparation to ensure that the research process can stand up to verification; in other words, if someone were to repeat the process, it could be replicated successfully.

There are six types of evidence sources:

- Documentation: e.g., letters, memos, and minutes of meetings.
- Archival records: e.g., organization charts, maps, and census data.
- Interviews
- Direct observation
- Participant observation: the observer is not someone who plays a passive role but who is actively involved.
- Physical artifacts: e.g., apparatus and tools.

The case in Sec. 9.2 was made up primarily of the distributed materials on the accounting system upgrade and information obtained through interviews. A list of the members present, a chart of the chain of command, and the accounting system software could also be considered as evidence. One does not necessarily need to conduct interviews or observations when writing a case. Cases can also be created based solely on documents and records.

Here, we refer to the difference between case studies and ethnography. In many of the fields where ethnographical approaches have been adopted, such as the field of education, it is difficult to control behavioral events and researchers are required to observe events as they occur. This is the same as in case studies, and the two methods share an awareness of issues and tools for analysis. However, case studies do not necessarily require direct and detailed observations of subjects, and the basis of discussion can be constructed with quantitative data.

The first principle of data collection is to utilize multiple sources of evidence. This is called triangulation of evidence. As triangulation in geometry enables us to determine the location of a point by measuring angles to it from known points at either end of a fixed baseline, the triangulation of evidence enables us to demonstrate a fact by combining multiple sources of evidence such as documentation, archival records, interviews, direct observations, participant observations, and physical artifacts. If respondents in an interview, for example, declare that informational meetings for users are a waste of time, one could evaluate the real meaning of the claims by filming the meetings and recording and analyzing participant behaviors and/or comparing the number of inquiries made to the accounting system department by people who attended or did not attend the sessions.

The second principle in collecting data is to build a case study database. This enables us to easily maintain and access the collected data, which enhances the reliability of case studies. Finally, the third principle in collecting data is to maintain the chain of evidence. This means that one should plan data collection so that it clarifies the path from data and evidence to conclusions.

After data have been collected, they are analyzed based on the evidence obtained. We will not go into methods of analysis here. After the data have been analyzed, the results are described as cases. Here, identities of the people in cases and reviews by the people involved should be considered. Disclosing the identities of the people in cases is extremely advantageous. This makes it easier for readers to recall related cases, and it also makes it easier to present appropriate evidence against a claim. However, maintaining anonymity in cases should be considered if the people involved could suffer loss by revealing their real identities. Advantages and disadvantages should be identified as to whether real identities should be disclosed or not. The case in Sect. 9.2, albeit fictional, uses people's names.

If the procedure for composing a case includes interviews of the people involved, the procedure should also include reviews of the draft by these same people. Although a detailed review makes it easier to detect errors and improves accuracy, the process can delay cases in being published. It is important to think about the time required for a review during the design stage of the case study. In the case in Sect. 9.2, the researcher would need to ask the appropriate person about any errors in analysis, and, if the case were to be published, about how detailed the description should be.

As previously discussed, the description in Sect. 9.2 from the case study point of view is just a portion of a case. A relevant case study would include a hypothesis, a discussion of relationships between the hypothesis and the data, and the author's conclusions.

Again, recall that we distinguished cases for research purposes (case studies) and those for teaching purposes (case methods) in Fig. 9.1. Moving over to the case method, one could create a teaching-purpose case based on the results obtained from a research-purpose case. As previously stated, the objective of a teaching-purpose case is to provoke discussion. A good case is characterized by six main points:

- The case has an educational theme appropriate to its purpose.
- It is well written (easy to read and interesting).
- It presents a problem to learners, and learners are able to recognize this.

I. Lead sentence of several lines creating a feeling for the whole story
II. Timeline to reach decision-making stage and key factors
III. Decision-making issues

The body of the case

IV. Appendix

Fig. 9.3 Standard organization for cases

- The case can be analyzed and discussed by learners.
- It allows learners to become decision-makers.
- It fosters discussion.

The standard organization for cases is shown in Fig. 9.3. The following is a possible description of the example in Sect. 9.2.

In the winter of 2004, A Telecommunication Corp. held an informational meeting about an accounting system upgrade via teleconferencing.

A Telecommunications is a leading manufacturer of routers, switches, and other communications equipment.

The company had over 1,000 employees, and maintenance of the internal information system was becoming an issue. The tax law was revised in 1998 and 2000 to correspond to the practical accounting procedural guidelines on software stipulated by the Financial Services Agency's Business Accounting Council. To comply with these revisions, A Telecommunications' accounting department added new functions to the system.

The success of these system modifications depended not only on the quality of the software, but also on user awareness. Mr. Kojima was the general advisor for the hardware development department at A Telecommunications.

After the above description, the author would like to continue with pertinent materials, questions to spur discussion, or instructor's notes.

9.6 Case Usage

Readers can acquire new knowledge by reading cases for research purposes as well as research papers. An important role of cases is also to facilitate communication. Cases can convey information to people in the field as well as researchers because they include descriptions of actual events. Thus, cases can mediate between researchers and people in the field. For example, if a researcher uses the technical term of "signaling" to explain the effect of greeting by the head of accounting, most non-researchers would find this difficult to understand, while the description in Sect. 9.2 would be understood by most people.

Cases for teaching purposes enable students to obtain the ability to make practical decisions. Another benefit of cases for teaching purposes is that students can get to know each other by sharing experiences and discussing cases. For example, it has been claimed that students discuss 500 cases before finishing the MBA program at the Harvard Business School. Such experiences lead students to build solid and life-long relationships.

Returning to the idea of communicating results and knowledge, let us compare the case and paper formats. Why do people read cases? This is because reading cases satisfies their intellectual curiosity. A more significant reason is that readers expect that reading cases will provide them with solutions to problems that they face or give them hints toward reaching solutions. However, it is extremely unlikely that situations described in cases and the situations readers face will be exactly the same. This is why papers need to be abstract and general.

Abstraction, or generalization, offers various benefits. Finding papers to read is a difficult task because there are huge numbers of these. Research communities mitigate this burden by selecting papers that are worth reading, i.e., only papers accepted through peer-reviews are published in the proceedings of international conferences and journals. This is possible because problems are abstracted to a certain extent so that lay people can understand the issues. Novelty is another key issue in the reading process by referees; evaluating the novelty of a paper is possible because abstraction helps distinguish research from other studies.

Selecting outstanding cases or clarifying the differences from previous cases, on the other hand, is a difficult task because cases are a description of concrete events. That is, the authors have to do the task of abstraction or generalization for research papers, while readers have to do the task of abstraction or generalization for cases. Thus, it is difficult to find cases that relate to one's problem. Even if cases were classified appropriately, researchers would end up by searching cases extremely similar to their own, and even though cases might be from different fields, researchers would have a difficult time finding knowledge that was invaluable to problem-solving.

The next point is ease of understanding, which is probably more a matter for the academic field than the difference between papers and cases. If someone is making a criticism of sociology, understanding one theory does not necessarily mean that he/she will understand another. However, economics tries to explain human behavior in terms of preferences, constraints, strategies, and equilibrium. This means that if we try to understand a theory, studying only different parts of theories already learned is sufficient. Because most case subjects do not go through the abstraction/generalization process, understanding one case does not mean one can easily understand another.

As explained above, while cases allow readers to experience something vicariously, they also include knowledge that is lost in the abstraction/generalization process. Therefore, cases are problematic in terms of searching and understanding to provide a means of communicating knowledge.

9.7 Cases in Field Informatics

This chapter explains cases by focusing attention on the possibility that they can be methods of conveying results and knowledge in field informatics apart from ordinal research papers. Cases are familiar in fields where case studies represent a major methodology, e.g., the study of law or medicine. Here, the question is what changes in existing case studies and case methods by introducing the viewpoint of field informatics.

First, the domain to apply case studies and case methods will be enlarged. This involves attempting to apply the methodology of case studies and case methods to the field of information processing, where the advantages of cases have yet to be fully acknowledged. As demonstrated by the substantial losses caused by a mistyped order at the Tokyo Stock Exchange in 2005, there is still a large gap between theory and practice in information system design. Employing the case study technique and recording data in case format are effective ways of approaching these issues more directly. In education, teachers in fields such as the management of technology (MOT) need to go beyond instructions on technology by creating a bond between the theories of intellectual property and financial affairs on the one hand and practical knowledge on the other. Case methods are well suited to such fields.

Second, more attention has been paid to the flow of information in the description of case studies. The case in Sect. 9.2 describes a decision-making process that focuses on the incentive of system users. This issue of incentive is difficult to resolve because information among participants is asymmetrical. This "asymmetry of information" results from the fact that the accounting department cannot directly observe whether or not their instructions are being incorporated into various departments. This concept, which is a central theme in information economics, has fostered a great deal of discussion. Economics, however, tends to place higher priority on simplifying analysis; thus, forcing phenomena and contexts apart often leads investigations off the track. In such cases, a good way to overcome such drawbacks is to write a case that focuses on the flow or asymmetry of information.

Third, distance learning systems and other information equipment can be used as a way of implementing case methods. One of the limits imposed on carrying out case methods is the number of students. As some students in an overcrowded classroom setting do not have the opportunity of making comments, teachers need to restrict the class size. Furthermore, when instructors use chalkboards or interrupt the flow of discussion in other ways, only one opinion can be heard. Consequently, it is impossible to hear multiple opinions at the same time. Although other people speak in distance learning systems, students do not have to worry about getting a chance to make comments. Furthermore, if an opinion can be classified automatically or semi-automatically and presented to the instructor and students, the distance learning system can help maintain the flow and perspective of the entire discussion. From the case study point of view, collecting difficult-to-gather evidence could also be made simpler by applying the advanced technology described in the chapter on human sensing. Although evidence collection may be easy, it is important to remember that one first needs to clarify problem awareness, as previously noted.

Thus, there are many components in the relationship between field informatics and case studies/case methods. One specific application of the case study format would be to design an information system. In the past, Universal Modeling Language (UML) has commonly been used for the design of information systems. A "use case" diagram, one of the figures used in UML, gives the required specifications for the system from the user point of view. Developing a use case often involves actual user behavior to be observed, a process that is quite similar to direct and participant observations in case studies. However, the objectives and attitudes are different. A case study is created to verify a hypothesis and produce findings that can be generalized. A use case diagram is described for system design. In a way, a case study could be thought of as something that provides perspective when a use case diagram is written. For that matter, system design does not end when design is complete—there is the need for constant improvements. In making these improvements, the relationship between the new system and people in the field would make a viable subject for a new case study.

The "case" form of expression might be problematic if we assume that one is working on the design of an information system. Cases for research-purposes require accuracy, which also means reading a great deal of text. Not only that, but reading one case does not always guarantee that another will be easy to comprehend. This might require a format that has abstraction and generality halfway between a case and a paper. In such cases, someone might be able to use a scenario format that extracts and describes important decision-making scenes from multiple cases. A discussion on this issue will be included in future work.

At the beginning of this chapter, we focused on the process of communicating results, saying that it is difficult to apply knowledge obtained from practice in the field to the existing paper-based mode of expression, especially in terms of demonstrating significance. This is not specific to the characteristics of the "field", as the same could be said for software, which has strong ties to information processing. The Japan Society for Software Science and Technology added a software paper category in April 2006. As ideas in software research have rarely been incorporated into practical software and very few papers have been published, researchers and developers have not often shared their ideas. The reason that so few papers have been published is that when authors write about their feelings, experiences, or details without any particular position, the paper is not viewed as a scientific description, but more as an essay or magazine article. However, if authors fumble around in an attempt to make their writing more abstract or general, the end result is a superficial explanation without any important interpretations. The arrangement of categories for software papers is an attempt to determine the value judgment of software; in other words, to create a scale to measure the value of software. Field informatics is also marked by many of the same problems. Researchers must continue working to formulate the groundwork for value judgements to generate more productive sharing of ideas.

In addition, we stated that knowledge is not always communicated from one researcher to another at the beginning of this chapter, but it can also be passed from researchers to practitioners. We believe that much of the discussion in this chapter will be useful to improve the quality of both communications from one researcher to another and from researchers to practitioners.

For first-time readers, the easiest way of understanding case writing is by actually reading cases. The literature in Yin (2008) contains many references to research-purpose cases. Readers can also refer to the literature in Garvin (2003) and Barnes et al. (1994) for cases with teaching purposes. After reviewing the relevant literature, use the flow described in this chapter to write a case about a familiar problem. Doing so should help you better understand the possibilities and challenges of case writing.

References

Garvin, D.A. *Making the Case: Professional Education for the World of Practice.* Harvard Magazine 106(1) 56–107, 2003

Yin, R.K. *Case Study Research: Design and Methods.* Sage Publications, Thousand Oaks, 2008

Barnes, L.B., Christensen, C.R., Hansen, A.J. *Teaching and the Case Method: Text, Cases, and Readings, 3rd* ed., Harvard Business Press, 1994

Chapter 10
Outreach Communication

Reiko Hishiyama

Abstract Citizens today are discovering that many critical issues remain to be solved, such as global climatic change, natural disaster response, resource and energy problems, and institutional fatigue of political and economic systems. Scientists, as experts, can make valuable contributions in given areas by providing technological solutions and an awareness of the latest technology. Creating new social systems and innovations that encourage sound social decision-making is also essential to ensure the sustainable preservation and use of natural environments, social infrastructures, and human resources, all of which took eons to develop and should be passed to the next generation. To tie scientists' contributions to the public's welfare in an organic manner, scientists must contribute to public-centered organizations tasked with establishing innovation and strengthening contact with local knowledge bases. To ensure that scientific knowledge and expertise are actually converted into societal value, it is further necessary to create both settings and frameworks in which a diverse set of actors, who will act as leaders in innovation, can interact and reach consensus. In this chapter, outreach communication is discussed as a feasible approach to tackling these problems. Originally, the word "outreach" meant "to exceed or go beyond" or "reaching out to the community", and it was mainly used in the public welfare field. It covers the welfare service activities that a community outreach worker, who normally works in a hospital or local health center, provides to those who need help or social support in external venues (Rowe 1999). However, "outreach" has recently been widely applied to the initiative activities that a researcher, who works in a public research organization or university, provides to or shares with the public or people outside the academic community. In science and technology fields especially, this word is used to express an attempt to promote public understanding of scientific knowledge

---

R. Hishiyama (✉)
Graduate School of Creative Science and Engineering, Waseda University, 3-4-1 Okubo, Shinjuku-ku, Tokyo 169-8555, Japan
e-mail: reiko@waseda.jp

T. Ishida (ed.), *Field Informatics*, DOI 10.1007/978-3-642-29006-0_10,
© Springer-Verlag Berlin Heidelberg 2012

and to delineate possible future scenarios with regard to science and technology for the next generation. These types of outreach activities are called "science communication" or "science and engineering communication." Lately, these words are widely used not only in the science and technology fields but also in the humanities and social sciences fields.

10.1 Feature of Outreach Communication

Present-day outreach, which includes technical communication, is undergoing a transformative shift away from one-way transmission of information by those in the scientific research community towards more interactive forms of communication.

Traditionally, outreach was a one-way activity, such as a public relations campaign or educational forum conducted by experts in the research community for citizens in the public sphere. Lectures, symposiums, and extension programs to increase the understanding of research activities are typical styles of outreach communication. In these still-popular styles, the expert speaks from a raised stage to citizens who form the audience. Although these styles are generally lively, they are still a form of one-way communication, so closing the gap between the expert and the citizen remains difficult. These communication styles also offer few opportunities for establishing flexible and interactive communication among citizens.

Citizens who live intimately with scientific or technological issues are beginning to take a strong interest in the social role of research and development, i.e., the future of social decision-making in countering the unexpected risks and problems that occur with the introduction of any new technology. For example, the explanations given by governments and the measures taken by experts in response to bovine spongiform encephalopathy (BSE) made apparent the communication problems to the public in the 1980s and 1990s. Against the backdrop of complex issues such as risk assessment in relation to scientific knowledge and food safety, the statements made and the activities suggested served only to increase the public's anxiety and heighten public mistrust of science and technology. This is often cited as an archetypal, real-world example of the challenges involved in communication between experts and the public (Irwin 2001). The lessons learned from such incidents have led to a greater focus on interactive communication, i.e., on the development of methods for creating environments within which diverse groups of people, including the general public, can discuss and debate matters of prime importance.

Science cafés, science workshops, and consensus conferences, all of which have been developed over the last few years, are communication methods that strongly focus on social interactivity. In these methods, to tackle the social issues that are associated with science and technology and which are intimately involved with the day-to-day lives of the public, both experts (not only scientists and technicians, but sociologists, economists and, at times, philosophers, poets and other experts, regardless of their field) and the public gather to share and discuss problems. The intention is to promote mutual understanding of the problems and acquire new perspectives. The process used should highlight the differences in values, differences in the ways of thinking about the problems, or differences in the direction of decision-making in future science and technology. These differences are brought about by the process through which science and technology are embedded into and accepted by society. At the same time, these communication methods show the social roles of science and technology, facilitate discussion to blaze new paths to a scientific future, and provide a chance to share mutually acceptable solutions to the problems (Gregory and Miller 2000).

From an expert's viewpoint, social interactivity brings numerous benefits such as obtaining new and valuable insights from citizens who have diverse local knowledge and experiences in different areas. Therefore, these communication methods offer new infrastructures to experts for knowledge production. The findings are fed back into the next research plan and provide a chance to shorten the distance between the experts and the public incrementally.

10.2 Styles of Outreach Communication

The outreach styles for science communications can be classified as either (1) outreach for publicity and promotion or (2) outreach for consensus building.

(1) Outreach for Publicity and Promotion

In outreach communication for publicity and promotion, experts widely provide and disseminate information concerning technical and scientific knowledge. This type of communication encourages the understanding of such knowledge and stimulates the exchange of ideas through the sharing of knowledge. In this communication style, communication planners emphasize how to transmit expertise to society or the community and ensure effective interactivity between citizens and experts. For example, astronomical knowledge is transmitted and expressed in many ways; it is not only associated with individual academic fields such as astrophysics and radio astronomy, but also with classical literature, art, and asterism, or calendars, since they are founded on astronomy. The most important thing is to ensure that the citizens and experts share the knowledge and take interdisciplinary perspectives on the issues beyond academic fields.

Social education facilities and outdoor activity sites are increasingly making use of science communicators' or interpreters' activities. These activities are intended to explain exhibitions or events using easy-to-understand words and phrases, which yield far better results than difficult technical jargon. At the same time, on-the-spot responses can be given to the public's doubts and questions. These communicators and interpreters can capture the public's ideas directly through dialogue and link these ideas to the experts.

Use of digital media as a means for the widespread sharing of information at the level of the public's perspective has also been actively trialed as a prominent and effective method of stimulating discussion and debate in the public sphere. Public outreach and education for NASA's Hubble Space Telescope is a successful example of the use of digital media in a communication channel (Christensen 2007). In particular, outreach activities that make full use of the great public space available on the Internet, and which give full consideration to interactive communication, are playing a major role as an interface through which experts can share their knowledge with the public.

(2) Outreach for Consensus Building

Outreach for consensus building is a communication method that enables the public and experts to think together about a sustainable society, and the future of science and technology. Specifically, it is meant for deliberation on critical issues, such as the validation of environmental conservation policies and safety assessments of nuclear power plants, for which predicting risk factors is difficult and for which building social consensus and acceptance is essential.

Prior attempts at outreach for consensus building included citizen-centered consensus meetings, citizen panels, and scenario workshops (Rowe and Frewer 2000; Joss 2002). Recently, 'public comments' have also become available on the Internet. The purpose of the public comment is to gather written opinions and comments from various parties. Consensus building is rooted in techniques for policymaking, in which the policymaker collects opinions and/or suggestions from the public or companies who will be directly affected by the issue. These opinions are then reflected upon in order to improve the process and principles by governing regulation. An extensive range of subjects are addressed by outreach for consensus building, from issues with wide-scale societal impact, such as genetic engineering and digital watermarking, to comparatively local-level issues that concern peoples' rights and interests, such as urban development and the protection of tidelands.

'Outreach for publicity and promotion' and 'outreach for consensus building' have a complementary and mutually reinforcing relationship. For the public to actively participate in shaping future decisions, a solid foundation for interaction that can encourage discussions dealing with high-level problems must be built. 'Outreach for publicity and promotion' contributes to building a basis for diverse interchange and provides the expertise and information necessary to enable the public to participate in decision-making. 'Outreach for consensus building' plays an important role in supporting the exchange of ideas for anyone who wants to participate in decision-making.

The communication difficulty that we are facing is that many recent issues and problems that are associated with science and technology are complex and require specialized expertise to be understood. Therefore, it is not easy for the citizens who do not have any particular scientific knowledge to obtain an overview of the issues and to make decisions for their future on the basis of their own knowledge. Furthermore, it must be acknowledged that the issues raised by the advance of science and technology often have indirect and long-term effects; they have far-reaching consequences and introduce uncertainty to business and daily life. Existing communication techniques are inadequate for discussing problems with these characteristics. This lack of an effective communication method by which the public can participate in future decision-making involving science and technology renders it impossible to create a foundation for the discussion and understanding of such problems. It then becomes difficult to weigh the various risks that may arise in the future, or to consider visions or proposals that may provide some relief.

To solve these problems, we need appropriate outreach communication techniques that enable the essentials of a complex problem to be transmitted from experts to citizens. A truly effective technique will bring benefits to many citizens by establishing a framework for dialogue with the experts and the sharing of opinions. It will also be necessary for citizens to be able to examine future overview scenarios, produce new knowledge from comprehensive viewpoints, and bring innovation to fruition. The following section explores this issue from the viewpoint of field informatics, i.e., outreach communication methods that incorporate communication tools.

10.3 Outreach Communication Methodology

10.3.1 Communication Models

Interactive communication by diverse participants including experts is the general target of outreach communication. For example, to discuss the risks of local pollution, we need participants who face the risks directly as well as those who are collaterally connected to the risks, such as people who are interested in the environmental movement or experts on the environment. To attain a more desirable social consensus, the discussions must incorporate the diverse perspectives of multiple stakeholders and citizens.

When classifying communication models from the perspective of sequential context, two types of communication can be distinguished: continuous sequential-type dialog communication, in which the basic sequence of a dialog follows a single context, and simultaneous parallel-type dialog communication, which permits dialog sequences to follow multiple contexts. The former is seen in the two-way communication conducted in environments such as a science café, where the roles are clearly established; the speaker presents topics, and the audience listens. The latter is exemplified by open and flat communication hierarchies in which roles are not clearly defined. The latter is more creative than the former and makes it easier for the speaker and audience to collaborate. In outreach communication, while the focus is on simultaneous parallel-type dialog, switching to the continuous sequential-type dialog is often necessary to increase the efficiency of information transfer. To aid in organizing such contexts, and to facilitate the exchange of information between experts and non-experts, a communicator or facilitator is often placed in the communication space.

10.3.2 Methods and Subjects of Communication

Communication planners prepare an interactive environment for addressing a communication theme in order to achieve effective outreach communication. The interactive environment can be implemented as a tool according to the aptitude of the participants, e.g., a game board, game cards, or games on computers. This makes knowledge sharing easy to understand from the public's standpoint.

As discussed in the previous section, problems that involve ethics or risk assessments are frequently dealt with in outreach communication. Let us take genetically modified (GM) foods as an example and think about how to deal with the risk and uncertainty of GM foods, the result of advances in the science and technology field of biotechnology. Food products using gene-recombination technology are thought of as contributing to the quality of food and increasing the food supply, while also helping to maintain public health and protect the environment. At the same time, many members of the public have deep-seated concerns about long-term safety issues, such as yet-unknown health-related side-effects or long-term effects on the ecosystem. This makes the social acceptance of GM foods difficult (Gaivoronskaia and Hvinden 2006). Some experts and citizens regard GM foods as having a high degree of usability and consider that safety and efficacy can be ensured in the same way as that used for existing foods, whereas others do not trust GM foods due to the concern that these foods have an unfamiliar origin, i.e., technology. In other words, some scientists are deeply concerned about potential negative effects on health or the environment.

We call such difference in opinion 'framing,' which also means the way to address the aspects of the issues. Different stakeholders often have different frames. With consideration of the differences among these frames, communication planners design the tool for interactive communication on the basis of the following two requirements.

The first requirement is to embed multiple viewpoints to produce the framing. This is meant to induce discussion and promote understanding of the issues among the participants and to encourage discussion about the overall issues and individual points. In embedding these themes, the need for ease-of-understanding while creating abstractions out of, and expressing, the basic structural characteristics of real-world issues must be carefully considered. This ease-of-understanding is generated by the appearance of a symbolic context while still providing the basic structural characteristics of real-world issues. That is, the outreach communication designer must prepare information that is universal, general, and simple enough to be understood without any kind of specialized, technical knowledge.

The second requirement is to prepare a flexible communication environment. Such an environment enables the restructuring or restating of the issue at hand

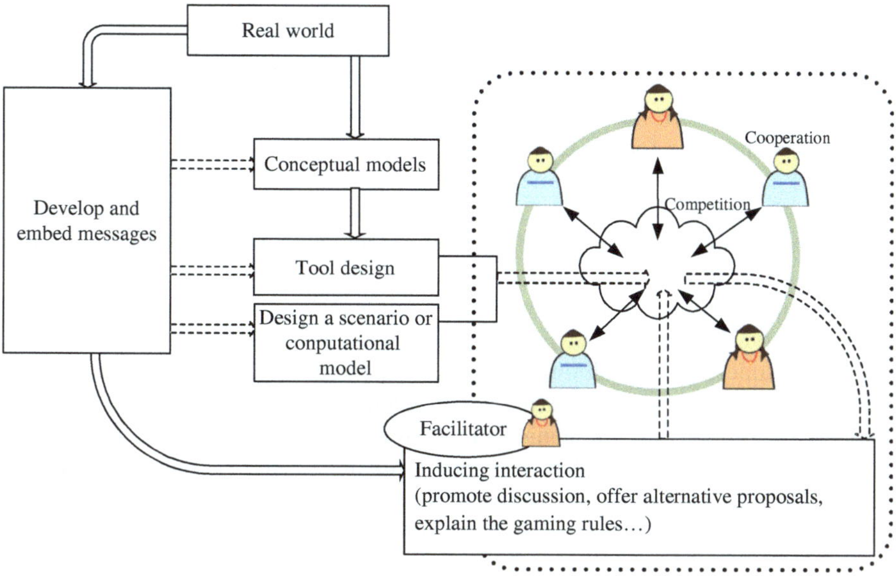

Fig. 10.1 Design process of communication tool

in response to the changes that are inevitable over time. Multiple scenarios or future projections may be presented dynamically as options corresponding to the changes in the participants' perspectives that will accompany progress in the discussion.

10.3.3 Building the Communication Environment

The design process treats the environment as a place of interaction and the content as a communication mechanism (Fig. 10.1). The first step for content design is to analyze the complex issues underlying the topic. On the basis of this analysis, the communication designer determines which facts must be shared, clarifies issues and specifies the scope targeted for consensus building, and delineates the specifics of the messages for the communication participants. The designer then creates abstractions out of the issues, which are simple enough to be understood considering the backgrounds of the participants. The designer then transforms abstractions into a conceptual model and sets about creating the specific requirements that the communication tools must satisfy. The tools are designed as a communicable model, if needed, by converting the conceptual model into a computational model. In this process, the input–output design as well as the computational procedure must be defined for the tool development if the tool will be actually run on a computer. Furthermore, the tools are designed on the assumption that the participants may be expected to have specific standpoints. Therefore, the tool designer should recognize the roles that have specific standpoints when designing the tools. Each role is expressed as an object-oriented model and requires much coordination and conflict/competition with other roles. Participants are expected to follow the assigned role under the constraints intended and specified by the designer. If the work has gone well up to this point, the communication theme, structure of the issue for discussion, and the roles to be played by the participants, should be established. At the same time, participants' social interaction as behavior is clearly designed in the form of the data and resources to be used and exchanged by the participants. The designer then finalizes the tools, taking into account what kinds of situations are likely to occur in accordance with how each role is played. For example, organic farmers, non-organic farmers (farmers of GM foods), multinational biotech companies, researchers, opposition citizen groups, consumers, food companies, public administration, etc. are regarded as stakeholders in the case of the gaming tool for GM food problems. The designer creates tools by making assumptions on the manner in which role interaction among stakeholders is embedded and about what problem structures can be shared among the participants and are linked to a decision-making process.

The tools constructed are only put into use after a preliminary trial. While overly complex tools can hinder comprehension of a topic, too much abstraction is not appropriate for an effective discussion. Therefore, the tools are adjusted in a trial run by the designer. In particular, adding or removing steps from the computational process, and refinement or simplification to tune the effects of expected events as communicative elements are based on the results of the trial run. If the tool runs on a computer, parameter adjustment and user-interface improvements may be required. Moreover, the designer evaluates the tool based on the actions, reactions, and statements of the participants in the trial run. The criteria are specified as follows: "whether all of the issues were adequately addressed as originally planned," "whether the communication gave rise to the anticipated levels of awareness and inquiry from the participants within the anticipated scenarios," "whether the communication was able to provide the participants with an impetus for deeper examination of the topics at hand," and "whether there was a sufficient degree of reality in relation to the evident issues."

A scenario is described in parallel with the development of the tools and is distributed as a handout before the outreach communication is started. The basic information underlying the issue is described in the scenario, and it also provides the common background and rules or procedures to the participants.

Additionally, the communication designer has to review the data collected in a debriefing. The debriefing is conducted as a short meeting at the end of the communication session, providing an opportunity to reflect on what happened and to recollect and organize the points discussed. The communicators or facilitators manage the time and the process of the communication, and encourage participants to be active in their roles, which include opposition or cooperation and to build consensus, while controlling the flow of the communication context as necessary. They also consider how best to bring to the surface the participants' local knowledge and make the most effective use of the prepared contents.

Appropriate contexts can be embedded in the tools in advance. One context sometimes triggers the generation of another context. The designer often designs and provides such a change of contexts as probabilistic events along with a measure of likelihood. In this way, they may represent unpredictable real-world phenomena, such as natural disasters, hazards brought about by the socially unjustifiable pursuit of profit, or contexts that can bring about ethical problems. For example, if a pollen-mediated gene flow, such as cross-pollination of genetic information between distinct plant populations due to insects or pollen dispersion, is regarded as one such event, the designer would incorporate it as an incident. The context is as follows: the organic farmer is accused of violating intellectual property rights regarding seeds and plants from multinational biotech companies when the genetic crossing is repeated more than a certain number of times. This design enables the provision of a context that represents the problematic side of genetic crossing. To bring about such a situation, the designer can, if using computer-based tools, incorporate phenomena as events that occur within specified parameters. The designer can also, if using playing-style cards as a tool, prepare in advance cards on which events are written. For example, the designer can place a "genetic-crossing event card" among several other kinds of event cards. This card, when drawn by a participant, provides an excuse for participants to increase or decrease their virtual assets and resources, or brings about incidents or accidents affecting all or certain participants. This pseudo-event will provide opportunities to offer opinions concerning the need for special cultivation techniques to secure the coexistence of GM crops and non-GM crops, the need for measures that prevent genetic crossing, and disclosure of information by biotech companies. For example, information on the location of field trials of biotech crops is important for organic planting systems because the biotech crops affect organic crops if the planting location is close. These problems of interspecific crossing may also offer administrative authorities a chance to establish opinions about a new regulatory framework.

Participants using tools designed following the above approach can experience multiple points of discussion against the backdrop of a dynamically developing context. In this way, all of the points involved in the discussion are designed to allow for reconstruction and redefinition in line with the context. In addition, discussion points are provided in such a way as to be used in common amidst the evolving situations and to guide participants towards solutions that are based on their experience in line with the overall contextual backdrop. The tools are not merely to provide the participants with opportunities for pseudo-experiences; they will also serve as a means to create new knowledge for problem solving.

Moreover, when designing tools, it is vital to maintain neutrality towards all of the participants while keeping in mind ethical considerations (Duke 1974). No matter what role the participants play, it is necessary to ensure that no disadvantages come about for any particular participant in the communication.

10.4 "Eco-experiment Game": A Practical Case Study in Environmental Issues

An outreach communication case study that uses personal computers connected to the Internet is introduced in this section. The communication theme of this practical case study is carbon circulation and global warming problems. The experiment was conducted by integrating participatory gaming with science communication (a science café). Participatory gaming was originally used for military tabletop exercises, was later applied to various fields such as business games and city simulations in social science, and now has become a useful practical research methodology in various academic fields (Hishiyama 2008). It enables participants who live in remote areas and have diverse local knowledge to establish outreach communication easily without the limitations of time or space. This section introduces a case study for putting communication into practice, including practicable steps for the construction of communication tools.

10.4.1 Problem Analysis

First, the designer should identify the messages that encourage communication. Although it is essential to take a global perspective of the future society when addressing the problem of global warming, local viewpoints are also needed to address local issues. Moreover, the solution is underspecified and the right answer is unknown. Against such a background, planners must make the existence of these problems both very tangible and widely shared, accumulate local knowledge, and create new solutions. The following points yield the intended results.

- Having explained the structural mechanisms of global warming from both natural science and sociological perspectives, the designer shares with the participants the issues that pose obstacles to achieving a sustainable society.
- To bring about sustainable economic activity while also protecting the environment, new systems, actions and methods to reduce carbon emissions are necessary to stop global warming. A notion of what kinds of options are possible, in terms of policymaking and concrete actions, can be acquired through the social context of activities mediated by communication tools or through the participants' perspectives on the conversation.

10.4.2 Designing the Communication Environment

In designing the communication tools, the designer must, having made the problem abstract, consider the following points: what kinds of information should be provided to the participants, with what timing and in what situations should the information be provided, and what unpredictable elements should be incorporated into the tools. In this case study, the problem is abstracted according to three elements: (1) time, (2) space, and (3) uncertainty and interaction.

(1) Time

The progress of global warming is visualized by the communication tools. This provides an opportunity for participants to give a sense of the time scale of global warming before consensus building. The time scale for global warming progression is incorporated within the tools, and as time progresses, a situation is created in which rehabilitating the environment is difficult because of the buildup of greenhouse gases over a long period of time. The individual economic activities of the participants affect the community as a whole, and those activities are associated with the progress of global warming. The tools demand that all participants act for mutual benefit with the goal of maintaining the global environment. They are designed to encourage the participants, each with varying circumstances and moving forward irreversibly through time, to make decisions individually while clarifying the societal dilemmas created when their pursuit of profit is at odds with the profit of society at large (i.e., stopping global warming).

(2) Space

The communication space represents a virtual world wherein heterogeneous virtual nations coexist as players. The designer should assess each nation's condition, assign local constraints to each accordingly, and design the activities and events considering the transboundary ramifications that might impact the nation conditions. The designer should also consider providing information that enables the exchange of status information.

Information managed at the national level includes information about each country's assets, resources, and budget situation. Even though these data are provided as open information, keeping them as closed information is possible if necessary. The status of the global economic situation and global environment are provided as global-level information.

(3) Uncertainty and Interaction

Events such as unexpected social hazards or natural disasters are assumed. In the early period, an event produces interaction and enables changes in the participants' conditions and the entire world's condition. It also provides a stimulus effect and new information to disrupt a monotonous context, and forces the participants to face unexpected problems. The occurrence of events is useful for reproducing circumstances in which a degree of uncertainty exists. However, depending on the likelihood of occurrence or the degree of the effect, events may sometimes produce outcomes not in line with the designer's intent. Therefore, caution is required since the degree of the effect will depend on each participant's condition.

Additionally, as an opportunity for participants to reach an awareness of the need to implement new solutions, the designer plans and provides opportunities for periodic, mock international global warming conferences (discussions) for the participants.

With the above in mind, the designer proceeds by fleshing out the design. Economic activities such as the sustainable use of forest resources and growth of agriculture and livestock, preconditions, and some constraints are set up for the nations with different values. The communication proceeds in rounds, and the designer builds into each round unexpected events, such as unpredictable social hazards or natural disasters triggered by global warming. These events are manifested stochastically in each round as necessary.

10.4.3 Preparation of Communication Tools

In accordance with the design process described above, the designer converts the communication design into tools that can easily be played, such as computerized gaming simulations including multiagent simulations. In the following case study, personal computers are used to establish outreach communication, but other environments such as game cards or game boards can be used. The communication tool should be selected according to the participants' ages, skills, and real-world concerns. The tool designer should check in a preliminary experiment whether there are any overlaps, contradictions, or errors in the game rules, or mismatch in the behaviors of nations. The designer makes minute adjustments, removing unnecessary rules while adding previously omitted rules. When the element of currency is to be introduced, the designer carefully examines the suitability of the scale of income and expenditures. The designer additionally confirms the consistency between the previously distributed handout on scenarios and the rules as put into actual practice. Furthermore, final adjustments are occasionally necessary in consideration of the anticipated ages and backgrounds of the participants. For example, rewriting the handout on a scenario in simpler vocabulary or preparing illustrations for elementary students is effective in assisting understanding.

By preparing and using the above-mentioned tools, the designer can create an environment in which the participants can virtually experience a real problem faced by the earth—the progress of global warming. In seeking a solution, the participants can also understand, through this experience, that the powerful greenhouse gas methane or human economic activities truly cause the greenhouse effect. The tools can also help lead participants to make concrete proposals based on the need for environmental policy options that are socially acceptable and that can be realistically put into practice without degrading the economy. The tools also provide a chance to understand the significance of undertaking governance on a global level and to consider the role that each nation should play.

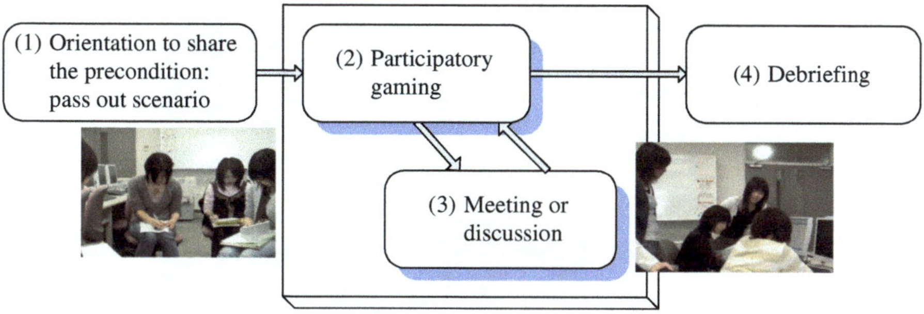

Fig. 10.2 Progress of communication exercise

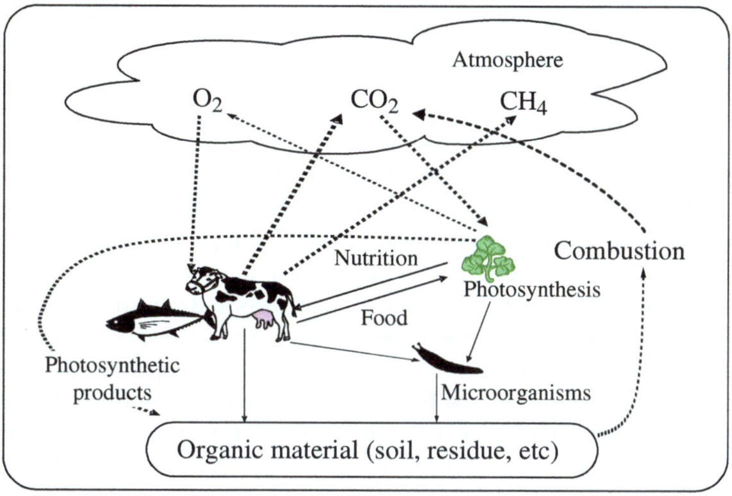

Fig. 10.3 Model of problem structure: carbon cycling and global warming

10.4.4 Practice in the Field

A general outline of the process of an actual communication exercise is illustrated in Fig. 10.2.

First, a scenario document entitled "Eco-experiment scenario: Let's protect our Earth from global warming!" was handed out before the exercise was conducted. The handout described the roles of carbon cycling and photosynthetic microorganisms in an easy-to-understand short story. To understand the global warming problem scientifically, the scientific mechanisms of carbon cycling and of utilizing renewable power sources such as solar, wind, and biomass must be understood. Making proper use of charts and illustrations in an easy-to-understand way is also beneficial. One illustration used in the scenario is shown in Fig. 10.3.

Rather than merely distributing the scenario materials, an expert was made available to explain the contents of the materials in plain words and to respond to

Fig. 10.4 Tetrahedral eco-dice linking simulated society to real society

any questions raised by the participants. Note that the handouts should, as necessary, include descriptions of the control-screen layout or the meanings of the input/output display icons when using computerized simulations, or explain the rules for usage and the functions of the various tools when using cards or game boards.

The participants then read the scenario, and the exercise was started. The participants played the role of government regulators and controlled the utilization volume of three energies: fossil fuel, natural gas (biogas), and biomass energy. To decrease the levels of carbon dioxide emissions, the participants planned to grow forest resources and rejuvenate the livestock industry, reconciling economic growth and environmental politics by collaborating with other participants. The participants could simulate the global warming problem and understood its social or scientific mechanisms.

Once every 4 years (expressed as a round number), all participants attended a virtual world conference to discuss the prevention of global warming. At this conference, they recognized and shared their views of global warming, discussed problem-solving interactively, made decisions regarding the targets for each country toward reducing greenhouse gases, and issued joint statements. At this point, the participants discussed matters such as requests for measures against countries producing excessive levels of greenhouse gases, requests for reducing methane emissions, which is considered to be more than 20 times as powerful a greenhouse gas as carbon dioxide, and the predicted effects of agricultural development on global warming.

A simple debriefing was conducted after at least ten rounds of the gaming simulation. In this debriefing session, with the aim of introducing the latest scientific and technological issues as they relate to climate change and global warming, the challenge of finding solutions to environmental problems by adopting biotechnology in conjunction with microorganisms was explained.

At the end of the exercise, the facilitator asked the participants to make small tetrahedral eco-dice (Fig. 10.4). The participants wrote selected actions to stop global

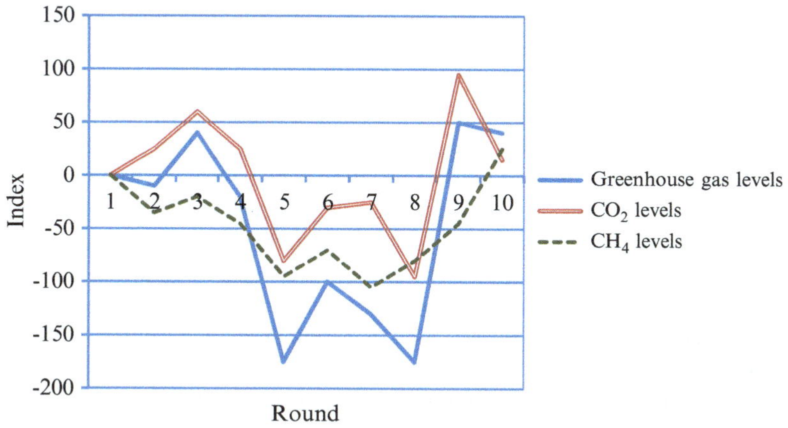

Fig. 10.5 Changes in global environment acquired from participatory gaming exercise

warming on each face of the dice to decide on workable actions in everyday life. The activities for stopping global warming written took the form of governmental policy declarations for the purpose of stopping global warming, as selected by each participant from among several options they deemed feasible. These tetrahedral dice served as a tool enabling participants to think about the interconnection between global problems and macro-scale living spaces. They also helped to encourage the participants to converse freely without constraints. The dice also gave the participants time to relax after the exercise and provided time to look back on the discussion.

The change in global warming over time in the exercise is illustrated in Fig. 10.5. The annual (per round) change in greenhouse gas levels achieved in each round is shown. In the early rounds, global warming advanced because of the positive use of energy. The simulated world global warming conference was held twice: after the fourth and the eighth rounds. In the conference held after the fourth round, an agreement was reached on a joint statement about the greenhouse gas emission target. After that, each participant made a reduction target and cooperated with other participants; therefore, the gas levels decreased after the fifth round. However, the cuts triggered a slump in economic activities. In the meeting held after the eighth round, participants reached agreement on a joint statement about the greenhouse gas emission target on a global scale. This decision, however, allowed each country to consider its own economic goal. As a result, the level of greenhouse gas rose again.

The opinions and comments gathered from the participants in this exercise broadly fell into two categories. One covered the hands-on acquisition of new information beyond the field of academic study, to not just the scientific field but also politics. First, the participants noted their great surprise in learning the scientific reality that methane, which is produced by biomass and is expected to serve as a natural resource, is a significantly more powerful greenhouse gas than carbon dioxide. This surprise was because carbon dioxide is conventionally thought to be a major cause of global warming. Moreover, from opinions concerning the fact that methane is primarily emitted as a natural by-product of agricultural activity, such as rice paddies and livestock, which is intimately related to economic activity, the exercise was ascertained to have encouraged a greater understanding of the structural makeup of the problem. Furthermore, this communication contributed to building the participant's foundation for understanding the advanced research findings, which were explained in the debriefing. The debriefing material described the findings of researchers with regard to the role of microorganisms: microorganisms that can directly use and consume methane exist, and a better understanding of their ecologies and roles could make possible the use of such microorganisms' capabilities in measures against global warming. This debriefing provided a chance to exchange opinions about the expectations for biotechnology and other technological developments.

The debriefing statements posed solutions that took into account the scales of time and space as mentioned in Sect. 10.4.2. This exercise presented participants with different policy options to match a world populated by countries with differing economic levels and different circumstances. The participants exchanged opinions on giving consideration to developing countries, which are most likely to be affected by environmental changes. They also mentioned that sustainable and balanced growth should be broadly shared within countries in any period of time, and that, to this end, providing incentives and constructing a framework in which countries can properly compete and cooperate is necessary. In addition, some participants said that any incentive system should consider the level of the target communities, i.e., community, city, or national level. They also discussed that human resource development and consumer education can be expected to be effective for countering global warming.

To tackle the serious problems faced by modern society, represented here by global warming, pluralist perspectives based on political and social views are required as well as scientific understanding. We can thus see that introducing communication tools is an effective first step in that the tools enable communication activity to provide participants with a comprehensive understanding of the issue. In addition, they provide diversified and multifaceted viewpoints that would be difficult to acquire, through simply discussion with experts in specific academic fields. However, the fixed problem area in the participatory gaming approach is a drawback: the participants may find it difficult to express their own local knowledge. For example, telecommuting technologies (which aim to reduce the movement of people and things) and sensor network technologies (which monitor and measure the global environment) can be considered valid for gathering knowledge and information through the use of information and communication technologies, particularly the World Wide Web technologies. There is no doubt that these technologies will make a large contribution to stop global warming. However, drawing out this solution by using the game introduced in this section would be difficult. Careful consideration must be given to how to extract the full potential of knowledge under the problem environment fixed by the gaming approach.

Moreover, tools that are developed for and run on computer systems can be quite costly in terms of skill requirements. In addition, the designer has to arrange the game by the communication theme. For easier gaming tool development, standardization of components remains a challenge for the future.

10.5 Possibilities and Challenges of Outreach Communication

Outreach communication provides an efficient framework to approach 'a can of worms' such as resource and energy depletion, i.e., problems that are complex and pluralistic and that nations are facing in the twenty first century. It also provides the opportunity to facilitate the participation and consensus of experts and citizens for building sustainable social systems. Outreach communication also holds the possibility of solving not just global-scale problems but also local and community problems, personal or mental problems, or even pressing day-to-day matters at the individual level—it can be used as a means of pursuing solutions to all types of problems that require attention at all levels. For example, an outreach communication tool that is collaboratively designed by university students and career development planners is provided for university students who are job hunting; the tool provides opportunities to think about careers from the student standpoint. Communication using this tool often births alternative stories told through the eyes of the students. These are completely unlike the job hunting or occupational counseling provided by the staff of career assistance centers in universities. In addition, there is the issue of how to form a harmonious, multicultural society in which the diverse citizens of various countries physically live in the same town. To address this, tools based on problem sharing amongst the citizens of various countries are currently being developed. These tools are intended to enhance a framework created to support multinational citizens with diverse needs. Recent outreach communication requires not only a focus on transmitting particular knowledge, but also the development of new methodology to offer entry into an understanding of the structure of complex problems and interactions. New methodology aims to collect opinions from citizens who have different local viewpoints and local knowledge, and to share the future opinions between the citizens and experts.

If providing a complete picture of a complex problem easily becomes possible by using outreach communication tools, more citizens will become actively involved in solving other critical problems or issues that are deeply and directly linked to their everyday life. They can consider several options for the serious problems that we are facing now and will face in the future, as an integral part of society. They can then work on these problems in conjunction with technical experts to reach social consensus through the sharing of citizens' perspectives.

The development of other various communication tools, which cover a wide range of themes, remains a challenge for the future. Facilitation skills, which encourage discussion among citizens who have various kinds of local knowledge, with or without expert knowledge, also should be explored with the aim of achieving outreach communication.

References

Christensen, L.L. *The Hands-on Guide to Science Communicators: A Step-by-step Approach to Public Outreach*. Springer, 2007

Duke, R. D. *Gaming: The Future's Language*. Sage Publications, 1974

Gaivoronskaia, G., Hvinden, B. *Consumers with Allergic Reaction to Food: Perception of and Response to Food Risk in General and Genetically Modified Food in Particular*. Science, Technology & Human Values, 31(6) 702–703, 2006

Gregory, J., Miller, S. *Science in Public-Communication, Culture, and Credibility*. Basic Books, New York, 2000

Hishiyama, R. *Participatory Gaming Simulation as a Science Communicator: Toward a Communication Bridge Between Specialists and Citizens*. World Congress on Engineering and Computer Science, 568–573, 2008

Irwin, A. *Constructing the Scientific Citizen: Science and Democracy in the Biosciences*. Public Understanding of Science (PUS) 10(1) 1–18, 2001

Joss, S. *Toward the Public Sphere—Reflections on the Development of Participatory Technology Assessment*. Bulletin of Science, Technology & Society 22(3) 220–231, 2002

Rowe, G., Frewer, L.J. *Public Participation Methods: A Framework for Evaluation*. Science, Technology & Human Values 25(1) 3–29, 2000

Rowe, M. *Crossing the Border: Encounters Between Homeless People and Outreach Workers*. University of California Press, Ewing, New Jersey, 1999

MIX
Papier aus verantwortungsvollen Quellen
Paper from responsible sources
FSC® C105338

If you have any concerns about our products,
you can contact us on
ProductSafety@springernature.com

In case Publisher is established outside the EU,
the EU authorized representative is:
Springer Nature Customer Service Center GmbH
Europaplatz 3, 69115 Heidelberg, Germany

Printed by Libri Plureos GmbH
in Hamburg, Germany